Numbers and Functions
Theory, Formulation, and Python Codes

Series in Computational Methods

Series Editor: Gui-Rong Liu *(University of Cincinnati, USA)*

Published

Series in Computational Methods
Volume 1

Numbers and Functions

Theory, Formulation, and Python Codes

G. R. Liu

University of Cincinnati, USA

World Scientific

NEW JERSEY · LONDON · SINGAPORE · BEIJING · SHANGHAI · HONG KONG · TAIPEI · CHENNAI · TOKYO

Published by

World Scientific Publishing Co. Pte. Ltd.

5 Toh Tuck Link, Singapore 596224

USA office: 27 Warren Street, Suite 401-402, Hackensack, NJ 07601

UK office: 57 Shelton Street, Covent Garden, London WC2H 9HE

Library of Congress Cataloging-in-Publication Data
Names: Liu, G. R. (Gui-Rong) author.
Title: Numbers and functions : theory, formulation, and Python codes /
   G.R. Liu, University of Cincinnati, USA.
Description: New Jersey : World Scientific, [2024] | Series: Series in computational methods ;
   volume 1 | Includes bibliographical references and index.
Identifiers: LCCN 2023057865 | ISBN 9789811287626 (hardcover) |
   ISBN 9789811287633 (ebook for institutions) | ISBN 9789811287640 (ebook for individuals)
Subjects: LCSH: Numbers, Complex. | Functions. | Python (Computer program language)
Classification: LCC QA255 .L58 2024 | DDC 512.7/88--dc23/eng/20240201
LC record available at https://lccn.loc.gov/2023057865

British Library Cataloguing-in-Publication Data
A catalogue record for this book is available from the British Library.

For any available supplementary material, please visit
https://www.worldscientific.com/worldscibooks/10.1142/13712#t=suppl

Desk Editors: Balasubramanian Shanmugam/Steven Patt

Typeset by Stallion Press
Email: enquiries@stallionpress.com

About the Author

Gui-Rong Liu received his Ph.D. from Tohoku University, Japan, in 1991. He was a Postdoctoral Fellow at Northwestern University, USA, from 1991 to 1993. He was a Professor at the National University of Singapore until 2010. He is currently a Professor at the Department of Aerospace Engineering and Engineering Mechanics, University of Cincinnati, USA. He was the Founder of the Association for Computational Mechanics (Singapore) (SACM) and served as the President of SACM until 2010. He served as the President of the Asia-Pacific Association for Computational Mechanics (APACM) (2010–2013) and an Executive Council Member of the International Association for Computational Mechanics (IACM) (2005–2010; 2020–2026). He authored a large number of journal papers and books, including two bestsellers, *Mesh Free Method: Moving Beyond the Finite Element Method* and *Smoothed Particle Hydrodynamics: A Meshfree Particle Methods*. He is the Editor-in-Chief of the *International Journal of Computational Methods* and served as an Associate Editor for *IPSE* and *MANO*. He is the recipient of numerous awards, including the Singapore Defence Technology Prize, NUS Outstanding University Researcher Award, NUS Best Teacher Award, APACM Computational Mechanics Award, JSME Computational Mechanics Award, ASME Ted Belytschko Applied Mechanics Award, Zienkiewicz Medal from APACM, AJCM Computational Mechanics Award, Humboldt Research Award, and SACM Medal from the Association of Computational Mechanics (Singapore). He has been listed as one among the world's top 1% most influential scientists (Highly Cited Researchers) by Thomson Reuters for a number of years.

Contents

Chapter 1

Introduction

Contents

1.1 Computational methods

This book is a part of the *Series in Computational Methods*. The book series aims to provide theoretically essential and practically important basic theory, formulation, and applications on computational methods. It covers general computational methods and techniques used in STEM education, as well as various fields in science and engineering. It will be written as an encyclopedia type of resource on basic theory, formulation, and codes on computational methods. Our readers need to have completed 9 years of education (primary and middle schools) to be able to study this book series starting from Volume 1 and then a few other elementary volumes of readers' choice. Experienced readers may jump to reading more advanced volumes.

Taking advantage of the fast development of computer hardware and software, our discussions will largely be accompanied by computer codes, so that the concepts, theories, and formulations can be immediately demonstrated using examples with results plotted. We will use primarily Python in the first handful of volumes on more fundamental topics and possibly other suitable languages for more advanced topics.

1.2 Why contribute to this book series

The Editor-in-Chief of this book series, Dr. Gui-Rong Liu, has been working in areas related to computational methods for over 40+ years. He developed his first finite element method (FEM) codes for nonlinear problems in 1980 and published more than 600 journal papers and 12 monographs in this area since then. After all these years of studying, using, and developing computational methods, he started to think about a means to help other interested individuals learn computational methods in a more effective, systematic, and smoother way. He has concluded that developing this book series is the best way to achieve this objective.

1.3 Two most fundamental concepts: Numbers and functions

This book is the first volume of the book series. It covers the two most fundamental concepts: **numbers** and **functions**.

A number is the basic building block for all objects used in computational methods, including functions, vectors, matrices, tensors. A function is the most essential device that converts numbers to numbers, and more generally, objects to objects.

Our discussion will be applications-focused without too much on theoretic proofs and overly lengthy derivations. Thus, much effort has been placed on coding to demonstrate the concepts, theory, and techniques. Python codes are used because they are currently, and in the foreseeable time, the most popular and easy-to-use tool. The idea is to break down all these difficult topics into simpler ones and conquere one by one immediately using the codes provided.

The materials of this book can be digested thorough different forms of learning processes, including classroom teaching, online courses, and, most importantly, self-study. As Python codes are provided, readers can easily see how the theory is formulated and how the solutions are obtained in terms of formulas, numbers, and graphs. Readers may also deepen their understanding via playing with codes and even further develop his/her own codes for solving other related problems.

In this book, we place emphases on closure properties of complex numbers and complex functions. This is because such properties are fundamentally important for the existence of solutions. In addition, computations all in complex numbers are readily feasible due to the increasingly improved performance of computer hardware and software. To ensure the existence of solutions and to avoid breakdown in computations, one can simply switch the computations all in complex numbers. Of course, if the solution does exist in the set real numbers and no breakdown in computations occurs, one should perform the computations in real numbers for better efficiency. Majority of the computations are currently done in real numbers.

The book is written in Jupyter Notebook format, so that description of theory, formulation, derivation, and coding can all be done in a unified document. This provides an environment for easy reading, exercise, practicing, and further exploration.

1.4 Who may read this book

The book is written for beginners interested in learning computational methods for solving problems in nature, engineering, and sciences. Readers need to only have at least 9 years (primary and middle schools) of education, including high school students, university students, graduate students, researchers, and professionals in any discipline. Engineers and practitioners may also find the book useful in establishing systematic concepts in computational methods.

1.5 Codes used in this book

The book is written using Jupyter Notebook with codes. Readers who purchased the book may contact the author directly at liugr100@gmail.com to request a softcopy of the book with codes (which may be updated) for free for academic use after registration. The conditions for using the book and codes developed by the author, in both hardcopy and softcopy, are as follows:

1. Users are entirely at their own risk using any part of the codes and techniques. The codes are written primarily for the proof of concepts and not necessarily for efficiency and robustness.
2. The book and codes are only for your own use. You are not allowed to further distribute without permission from the author.

3. There will be no user support.
4. Proper reference and acknowledgment must be given for the use of the book, codes, ideas, and techniques.

These codes are often run with various packages/modules. Therefore, care is needed when using these codes because the behavior of the codes often depends on the versions of Python and all these packages/modules. When the code does not run as expected, version mismatch could be one of the problems. At the time of the writing of this book, the versions of Python and some of the packages/modules are as follows:

- Python 3.9.16 :: Anaconda, Inc.
- Jupyter Notebook 6.1.5

When issues are encountered running a code, readers may need to check the versions of the packages/modules used. If Anaconda Navigator is used, the versions of all these packages/modules installed with the Python environment will be listed when the Python environment is highlighted. You can also check the versions of a package in a code cell of the Jupyter Notebook. For example, to check the version of the current environment of Python, one may use the following:

```
1  !python -V          #! is used to execute an external command
```

Python 3.9.16

```
1  !jupyter notebook --version
```

6.1.5

If the version is indeed an issue, one would need to either modify the code to fit the version or install the correct version in one's system. It is very useful to query on the web using the error message, and solutions or leads can often be found. Online AI tools, such a ChatGPT, Bard, and Bing, can also be quite helpful. This is the approach the author often takes when having an issue in running a code.

This book will not discuss how to use Python. There is plenty of litera-
ture openly available online this topic. Interested readers may also refer to
Chapter 2 in Ref. [1] for a concise description on using Python. As Python
is quite easy to learn, one may simply use the codes and examples given
in this book and jump start to learn Python in the process of studying the
technical subjects of this book.

1.6 Use of external modules or dependencies

To use Python and codes provided in this volume, we import necessary
modules and functions. The following are the most essential ones:

```
1  import sys                                    # import "sys" module
2
3  sys.path.append('../grbin')                   # Relative directory
4                      # Or absolute folder like 'F:\\xxx\\...\\code
5
6  #Author's own grcodes module is placed in folder grbin
7  import grcodes as gr
```

To view the codes in the imported module, one may just using the following code by uncommenting it (It will produce a long output):

```
1  import inspect
2  #source_code = inspect.getsource(gr)    # to view everything in gr module
3  source_code = inspect.getsource(gr.cheby_T) # to view any function in gr
4  print(source_code)
```

```
def cheby_T(n, x):
    '''Generate the first kind Chebyshev polyncmials of degree (n-1) '''
    if   n == 0: return sp.S.One
    elif n == 1: return x
    else:        return (2*x*cheby_T(n-1,x)-cheby_T(n-2,x)).expand()
```

Alternatively, one may use any text editor to view and change the codes.
To use any code function, say printx() in the imported module gr, for example, use gr.printx. The following is an example:

```
1 x = 8
2 gr.printx('x') # when gr. is used, the code function is from the grcords
```

x = 8

To avoid frequent importing of lengthy external modules, we put all the frequently used modules in the "commonImports.py" file as follows:

```
1  from __future__ import print_function
2  import numpy as np                              # for numerical computation
3  import sympy as sp                          # sympy module for computation
4  import numpy.linalg as lg                    # numpy linear algebra module
5  import scipy.linalg as sg                     # scipy linear algebra module
6  import scipy.integrate as si
7  from scipy.stats import ortho_group
8  import importlib
9  import itertools
10 import inspect
11 import csv
12 import pandas as pd
13
14 from grcodes import drawArrow, plotfig                        # for plots
15
16 import math as ma
17 from sympy import sin, cos, symbols, S, lambdify, init_printing
18 from sympy import pi, Matrix, sqrt, oo, integrate, diff, Derivative
19 from sympy import MatrixSymbol, simplify, nsimplify
20 from sympy import factor, expand, nsimplify, Matrix, ordered, hessian
21 from sympy.plotting import plot as splt
22 init_printing(use_unicode=True) # for latex-like qualityprinting formate
23
24 from matplotlib.ticker import MultipleLocator
25 import matplotlib.pyplot as plt                     # for plotting figures
26 import matplotlib as mpl
```

In the beginning of each Jupyter Notebook (chapter), we simply import everything in commonImports:

```
1 #Often used external modules are in commonImports placed in folder grbin
2 from commonImports import *
```

In general, importing the same module multiple times does no harm. In fact, Python ignores all the latter importations if it is already in the cache. Due to this, if one has made changes to the imported module, reloading the module is required to have the modification take effect. This is done using importlib.reload(). For example,

```
1  # grcodes was imported as gr, reload it when grcodes is modified
2  importlib.reload(gr)
```

1.7 Use of help()

To find more details on what a module is or what an object in a module does, use help() after importing. For example,

```
1  help(gr.cheby_T)
```

```
Help on function cheby_T in module grcodes:

cheby_T(n, x)
    Generate the first kind Chebyshev polynomials of degree (n-1)
```

Note that in the code cells given in the book, necessary comments (starting with "#") are used to provide additional explanations. These comments are put on the right-hand side in the cells so as to not disturb the reading of the code much.

Reference

[1]   Liu GR, *Machine Learning with Python: Theory and Applications*, World Scientific, 2023.

Chapter 2

Real Numbers

Contents

Discussions in this book are accompanied with Python demonstrations. We thus import the major necessary Python modules (called dependencies) at the beginning of each Jupyter Notebook (a chapter) by pressing together "Ctrl+Enter" so that the codes can work properly without the need to import modules in each of the code cells. If an error message like "name 'xxxx' is not defined" is encountered, readers may jump back to the following cell and execute it. If a module cannot be found, readers may use "!pip install 'module_name'" and then execute the following cell again.

```
1  # Place curse in this cell, and press Ctrl+Enter to import dependences.
2  import sys                          # for accessing the computer system
3  sys.path.append('../grbin/')  # Change to the directory in your system
4
5  from commonImports import *        # Import dependences from '../grbin/'
6  import grcodes as gr                # Import the module of the author
7  importlib.reload(gr)                # When grcodes is modified, reload it
```

A number is the most basic element in computational methods. It is the smallest building block for all the objects (functions, vectors, matrices, tensors, etc.) used in algebraic operations and functional analysis. It is thus of fundamental importance in both algebra and calculus, and beyond. A member can be real and also complex. This chapter discusses the property and behavior of real numbers and the essential operations to real numbers used in computational methods for problems in science and engineering.

Many readers may already be familiar with real numbers since it is used all the time. Due to this "familiarity", one may often overlook its behavior. A systematic examination of numbers can be a good exercise for theoretical understanding, formulation, and coding. The simple fact is when a computation breaks down, it is often rooted in an operation on a number. Also, when a computation finds a solution successfully, it may also be because of the behavior of the numbers.

This chapter is written in reference to the documentation on related Python, Numpy, and Sympy and other relevant wiki pages, including Field and Real numbers. "Discussions" with ChatGPT, Bing, and Bard together with the Google searched pages have also helped in the preparation of the chapter.

2.1 The set of real number $\mathbb{R}$

Generally speaking, a real number is used to measure the value of a quantity, such as mass, length, time, speed, and temperature. Real numbers include three different types of numbers:

- all integers, such as -2, 0, 1, 2, and 5, denoted as set $\mathbb{Z}$, in which 0 is particularity special and is a must-have when a set becomes a "vector space",
- all rational numbers that are fractions of integers, such as $1/2$, $-1/2$, $-3/2$, and $3/5$, denoted as set $\mathbb{Q}$,

- all irrational numbers, such as $\pi \approx 3.142$, $e \approx 2.718$, and $\phi = \frac{1+\sqrt{5}}{2} \approx 1.618$ (the golden ratio), $\sqrt{2}, \sqrt{3}, \sqrt{5}, \ldots$, just to name a few.

All these numbers can be positive or negative, filling the entire one-dimensional (1D) real coordinate space, and are all bounded between $-\infty$ and ∞. The entire set of real numbers is denoted as $\mathbb{R}$. Figure 2.1 shows schematically some real numbers in a 1D domain in $\mathbb{R}$.

A former definition of real number is by the Dedekind cut of rational numbers. Based on this definition, real number is **continuous** in $\mathbb{R}$.

Clearly, $\mathbb{R}$ is ordered because all the real numbers can be ranked based on their values. Its extensions will be to functions, vectors, matrices, and many other objects.

The set of real numbers $\mathbb{R}$ is an ordered field denoted as $\mathbb{F}$ that is a fundamental algebraic structure equipped with the **arithmetic operations**:

- addition $+$,
- subtraction $-$, which is the inverse addition, meaning that for any number $a - a = a + (-a) = 0$,
- multiplication $\times$ (often omitted in expression),
- defined-division $\div$ (or $/$ or $-$ or sometimes $:$) that forbids the denominate to be 0.

These arithmetic operations are allowed to have maneuver properties, including **commutativity**, **associativity**, and **distributivity** for addition, which means the following:

- **Commutativity**: For any two numbers a and b: $a+b = b+a$; $a \times b = b \times a$.
- **Associativity**: For any three numbers a, b, and c: $(a+b)+c = a+(b+c)$.
- **Distributivity**: For any three numbers a, b, and c: $a \times (b+c) = a \times c + a \times c$.

This set of conditions is also known as **axioms** for the set of real numbers.

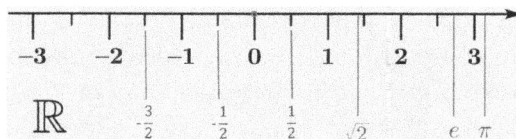

Figure 2.1. A 1D domain where infinite number of real numbers live. Image modified based on the one from commons.wikimedia, the free media repository, public domain, by Phrood~commonswiki.

Basic operators are built into a computer at the machine language level to perform these four corresponding arithmetic operations. Other more complicated operations for numerical computation are essentially carried out using algorithms consisting of these basic operators.

2.2 Limits and accuracy of a real number in Python

In Python, real numbers include numbers in integer types and float types. In actual computations in a computer, there is always a limit on the number of bits, and hence, the value of any number has a finite bound. Thus, there is possibility for the value of a newly produced member after some operations exceeding the bound. This is when we encounter the so-called **overflow or underflow**, leading to breakdown in the computation.

2.2.1 *Limits for integers in Python*

Due to its exactness in digital representation in computers (within its bits limit), integers (defined as integer type) in a computer are often used for precise indexing, ordering, and ranking. The information on integers in Python depends on the computer system used. One may find the information in his/her computer using the following:

```
1  integer_max = sys.maxsize
2  print(f" The largest integer in this computer = {integer_max:,}")
3  print(f" The largest integer using 64 bits    = {2**63:,}")
```

```
The largest integer in this computer = 9,223,372,036,854,775,807
The largest integer using 64 bits    = 9,223,372,036,854,775,808
```

As shown, the largest integer is larger than 9 quintillion (10^{18}) because the author's laptop runs a 64-bit system, in which 63 bits are for the value and 1 bit is for the sign. The upper limit of an integer is $\sim$9 quintillion, and the lower limit is ~ -9 quintillion.

Just to give an idea on how big the largest integer is, the population of the whole world in 2022 is about 8 billion (10^9). Thus, it is unlikely one will hit this limit in computations in solving usual practical problems if the problem is well posed and the formulation is done properly. Therefore, if an over- or underflow is encountered, chances are the causal are just some code bugs or an improper setting of the problem.

When Numpy module is used, the information on an integer type can be found as follows:

```python
1  int64_bits  = np.iinfo(np.int64).bits  # bits occupied by int64 (default)
2  int64_upper = np.iinfo(np.int64).max       # Largest representable int64
3  int64_lower = np.iinfo(np.int64).min               # -max
4
5  print(f"The number of bits occupied by a np.int64: {int64_bits:,}")
6  print(f"The upper limit of a np.int64: {int64_upper:,}")
7  print(f"The lower limit of a np.int64:{int64_lower:,}")
```

```
The number of bits occupied by a np.int64: 64
The upper limit of a np.int64: 9,223,372,036,854,775,807
The lower limit of a np.int64:-9,223,372,036,854,775,808
```

We get the same integer limit as we obtained from the system, as expected.

2.2.2 *Limits and accuracy for float real numbers*

For large-scale computations, Numpy is often preferred. In Numpy, the bits, bounds, and precision information for a float type can be found using the following code (one may find more using "help(np.finfo)"):

```python
1  float_bits  = np.finfo(float).bits          # bits occupied by a float
2  float_eps   = np.finfo(float).eps       # gap between floats (precision)
3  float_upper = np.finfo(float).max       # Largest representable number
4  float_lower = np.finfo(float).min               # the -max
5
6  print("The number of bits occupied by a float  :", float_bits)
7  print("The precision of presentation of a float):", float_eps)
8  print("The upper limit of float: ", float_upper)
9  print("The lower limit of float:", float_lower)
```

```
The number of bits occupied by a float  : 64
The precision of presentation of a float): 2.220446049250313e-16
The upper limit of float:   1.7976931348623157e+308
The lower limit of float: -1.7976931348623157e+308
```

As shown, the domain of the floats is extremely large. These upper and lower limits are $-\infty$ and ∞ in the computer. The precision of a float is $\sim$2.22e$-$16, which is also called **machine accuracy** or **machine epsilon**. This means that a new float number generated has more than $\sim$16 decimals, beyond which it is rounded off. Therefore, real numbers represented in a computer are not really continuous. There are very tiny gaps. As these gaps are so small, it does not affect practical uses. If it is found affecting the results, one may use float128 type if the computer platform supports it. If it is still not enough, one may resort to some arbitrary precision libraries such as gmpy2.

In conclusion, in most of our practical computations in a modern computer, we can safely treat real numbers being continuous and bounded by infinities, provided the problem is well-posed, and the formulation is proper. We will see some examples later about ill-posedness and good practices in formulations and computations.

2.3 Examples: Real numbers produced using Python

The following Numpy code generates randomly N real numbers within a given domain bounded by real numbers a and b. We first define the boundaries in float type in Python:

```python
1  N = 8                        # integer, the number of floats to be created
2  a = -10.; b = 10.            # floats, domain [a, b] hosts these numbers
3
4  print(f'N = {N}; the data type of N: {type(N)}')
5  print(f'a={a}; the data type of a: {type(a)}')
```

```
N = 8; the data type of N: <class 'int'>
a=-10.0; the data type of a: <class 'float'>
```

Note that in Python, there is no need to specifically define the data type. The assignment "=" itself is used to define it: if the assigned number is not followed by or containing a ".", it is an integer, otherwise a float.

Next, generate a list of real numbers, and put all those in Python "List" that is a container of objects including numbers:

```python
1  real_number=[np.random.uniform(a, b) for i in range(N)]      # in a list
2  print(f'A randomly generated real number:')
3  print(real_number[:3])                    # print out the first 3 real numbers
4  print(f'The type of this real number: {type(real_number[0])}')
```

```
A randomly generated real number:
[-5.501680894241807, -9.18578591485529, -3.9025747015130037]
The type of this real number: <class 'float'>
```

Printing out long lists of double-precision floats will occupy significant space. It can also be difficult to view. Thus, we often use gr.printl() in our Jupyter Notebook to print out such lists. Note that it does not change the precision of the float inside the memory of the computer.

```
1  gr.printl(real_number, 'Shortened floats:\n', n_d=4) #     with 5 digits
```

```
Shortened floats:
[4.5339, -3.7835, -6.453, -5.7827, -8.0524, 9.4134, 2.9119, -6.7205]
```

If needed, one may also sort the list in order of the values, using

```
1  real_number.sort()      # sort the list using Python built method sort()
2  gr.printl(real_number, 'Sorted list:\n', n_d=4)
```

```
Sorted list:
[-8.0524, -6.7205, -6.453, -5.7827, -3.7835, 2.9119,  4.5339, 9.4134]
```

Python is a so-called object-oriented computer language. When the float object (real_number) is created, many code functions (also called methods) may be used using simply ".". The sort() used in the above code cell is one of those.

To generate Numpy array, use the following:

```
1  np.set_printoptions(precision=3)      # control print out digits in numpy
2  real_number = np.random.uniform(a, b, size=(N,))          # a numpy array
3  real_number                 # at last line in the code cell, it is printed
```

```
array([-1.537, -7.767, -0.193, -1.472,  4.94 , -9.113,  5.02 3,  9.247])
```

Note here that "real_number" is now reassigned. It comes with a np.array object. To sort the array, use the np.sort() method:

```
1  real_number = np.sort(real_number)     # sort the array via a numpy method
2  real_number
```

```
array([-9.113, -7.767, -1.537, -1.472, -0.193,  4.94 ,  5.02 3,  9.247])
```

Readers may pay attention to the subtle difference between list and Numpy arrays. One may also refer to Chapter 2 in Ref. [1] for a concise description on using Python.

One may use a large N to have a given domain of $[a, b]$ covered with real numbers as dense as one wishes to (up to machine accuracy and memory). This may help establish some feeling that the real numbers cover $\mathbb{R}$ continuously.

Special and often-used irrational numbers can also be easily accessed or generated in Numpy:

```python
1  print(f'π≈{np.pi}, e≈{np.e}, φ≈{(1+np.sqrt(5))/2}')
```

$\pi \approx 3.141592653589793$, e $\approx 2.718281828459045$, $\varphi \approx 1.618033988749895$

Note that irrational numbers have infinite number of digits after the decimal place. Thus, such a number can only be approximated in a computer.

2.4 Random sampling of numbers over a domain

In computational methods, one often needs to sample numbers in a random manner, as we just did in the previous example. In Python's standard random module, methods can be conveniently used and "simulated" random numbers can be generated in a given domain.

Let us produce random integers, using random.randint(), which samples an integer uniformly in a given range of integers:

```python
1  import random                        # standard Python built-in module
2  #help(random.random)             # help() are often used to find more details
```

```python
1  na, nb, n = 1, 100, 5                        # n integers in na~nb
2
3  for i in range(n):              # Use a loop to generate n random integers
4      print(np.random.randint(na, nb),' ',end ='')        # in na~nb
5
6  print('\n')
7  for i in range(n):                            # Generate again
8      print(np.random.randint(na, nb),' ',end ='')    # n random integers
```

46   35   74   98   58

99   40   25   24   78

One may execute the above code cell for multiple times (by pressing Ctrl+Enter). Each time, it generates five random integers twice. These numbers generated are "random" because the generated numbers are different each time.

2.5 Controlled random sampling

Now, let us redo the same but use random.seed() to specify a seed for each generation:

```python
np.random.seed(8)      # seed 8 (any integer) for random number generation
for i in range(n):                         # Generate n random integers
    print(np.random.randint(na, nb),'  ',end ='')

print('\n')                                # print out an empty line
np.random.seed(1)                 # The same seed value (try also seed(2))
for i in range(n):
    print(np.random.randint(na, nb),'  ',end ='')
```

```
12   3   68   58   88

38   13   73   10   76
```

This time, the same set of numbers is generated, which is a kind of **controlled random sampling** by a seed value. The function random.seed() is used to ensure the repeatability when one reruns the code again, which is important for code development to ensure reproducibility. It is quite frequently used in computational methods and also in this book.

We see the fact that random numbers generated by a computer are not entirely random and controllable to a certain degree. This pseudo-random feature is useful: when we study a probability event, we make use of the randomness of random.randint() or random.random(). When we want our study and code to be repeatable, we make use of random.seed() so that when others use the code, it gives the same behavior.

Note that the seed value can be changed to any other number, and a different seed value gives different random numbers.

Let us now generate real numbers in a controlled random manner:

```python
np.random.seed(7)                     # seed for random number generation
n = 5
for i in range(n):                        # generates n random real numbers
    print(np.random.random())
```

```
0.07630828937395717
0.7799187922401146
0.4384092314408935
0.7234651778309412
0.9779895119966027
```

These real numbers generated are in between 0 and 1. Each number is produced by generating a random integer first using random.randint() and then dividing it by its maximum range. The reader may switch on and off random.seed() or change the seed value to see the difference.

In addition to the uniform distribution used above, one may use normal, binomial, Poisson, exponential, gamma, beta, and standard_caushy and chisquare in a similar manner. For example, the following generates random real numbers with a normal (also called Gaussian) distribution:

```
1  np.random.normal(.0, 5., size=(7,))    # using np.random, gives an array
```

```
array([ 0.171, -2.517,  4.496, -1.613,  3.481, -5.804, -4.357])
```

The most frequently used normal distribution may be the standard normal distribution that has 0 mean and the standard deviation is 1.0:

```
1  np.random.normal(size=(7,))  # standard normal distribution, the default
```

```
array([-0.881, -0.71 ,  0.15 , -1.169,  1.602,  1.857,  0.759])
```

Apart from the major use of random numbers in statistical and probability analysis, random numbers can often be used to numerically "prove" theories. If a theory is tested true via a sufficient number of random tests, it is likely true. It is not a replacement for theoretical proof but can be a good demonstration for establishing some intuition, which can be a good help in understanding the theoretical proofs. We will use this approach quite frequently in this book.

2.6 Concept of closure

The term "closure" is a very important concept in examining the property of a set, for example, the set of real numbers $\mathbb{R}$ under discussion in this chapter. When a set is said to be closed under an operation, it means that when arbitrary elements in the set are subjected to the operation, the resulting new elements must always be within the set.

For example, the set of integers $\mathbb{Z}$ is closed under the addition operation. This is because adding up any two integers $\mathbb{Z}$, the result is always an integer that is also in $\mathbb{Z}$. In particular, adding $3 + 5$ results in 8 that is still an integer, etc.

A counterexample is that the set of integers $\mathbb{Z}$ is not closed under the defined division operation because by dividing an integer by another non-zero integer (all in $\mathbb{Z}$), the result is not necessarily in $\mathbb{Z}$. In particular, $3/2$ results in a rational number that is not in $\mathbb{Z}$.

The significance of the closure concept is that it guarantees consistency of the operation: the new result from the operation will always stay in and never "disappear" from the set. This enables further necessary operations to all the elements in the set, and nothing will leave the set.

For our study in this chapter, we argue that the set of real numbers $\mathbb{R}$ is closed under all the arithmetic operations: the results of any of the arithmetic operations to any elements in $\mathbb{R}$ will always be in $\mathbb{R}$.

Let us write the following code to numerically "prove" this argument.

2.7 Numpy code for the closure test of real numbers

We write the following Numpy code for numerical tests on the closure of real numbers (floats). These numbers are generated randomly based on either the uniform or normal distribution:

```
 1  def closure_test(N, uniform=True):
 2
 3      '''Perform numerical tests on the closure of real numbers
 4         under arithmetic operations. Numbers are randomly generated.
 5      Input:
 6         N: number of tests.
 7         uniform: if True, sample floats using uniform distribution.
 8                  otherwise, using the standard normal distribution.
 9      Output: closed or not closed
10      return: None
11      '''
12      np.random.seed(None)                              #  set seed off
13
14      if uniform:
15          random_float = np.random.random
16      else:
17          random_float = np.random.normal
18
```

```
19    record = 0
20    fnt = np.isfinite
21    for i in range(N+1):        #generates N+1 random real numbers and test
22        a = random_float()
23        b = random_float()
24        apb = a + b
25        amb = a - b
26        atb = a * b
27        adb = a / b
28        if (not (fnt(apb) or fnt(amb) or fnt(atb) or fnt(adb))):
29            record = i
30            print(f"Floats real numbers is not closed!")
31
32    if (record == 0): print(f"Float real numbers are closed!")
```

Let us run the test for 10 million times using uniform sampling. This takes about 1–2 minutes to run on the author's laptop:

```
1  N = 10_000_000
2  closure_test(N, uniform=True)
```

Float real numbers are closed!

The results are true. Let us run the test again by sampling the floats using the standard normal distribution. This also takes about 1–2 minutes:

```
1  N = 10_000_000
2  closure_test(N, uniform=False)
```

Float real numbers are closed!

The results are again true. The readers may increase the number of tests to gain some conviction or find otherwise. The running time should be roughly linear with N.

Clearly, if one would purposely use very big float numbers that are close to the maximum float number, or keep adding up large positive numbers many times, or simply multiply two huge numbers, the test will fail because of the limits of floats in the computer. Given in the following is an example:

```
1  1.e155 * 1.e155
```

We got an infinity number (overflow). Obviously, this is a case made up for failure. One can also make up for failure by using a loop to add up a huge number of even small numbers. In real life, it is unlikely to happen. In addition, we have techniques to prevent this kind of big numbers from getting into the computation process.

2.8 Normalization or scaling

Note that the uniform and the standard normal distributions are used in the numerical tests given above. This means that the float numbers generated are within $[-1, 1]$ or around 0. This limits the growth of the new float number undergoing the arithmetic operations.

We thus give the following remark:

> When performing computations with float numbers, normalization of the numbers is important to control the growth.

The often-used techniques for normalizing datasets of numbers are as follows:

1. the **min–max scaling** that brings the value data points in the dataset within $[-1, 1]$ or in a small finite range defined by the user;
2. the **standard scaling** that uses the standard normal distribution; in this case, the value of the data points may be beyond $[-1, 1]$ but will be centered at 0 and quite close to 0.

Such a scaling is done to the data before computation. After the computation is done, one can scale back. Readers may find more details on these two techniques in Chapter 3 of Ref. [1].

Also, it is a good practice to perform normalization at the formulation stage, as done in Refs. [2, 4, 5]. One can simply divide both sides of an equation with the largest number involved in the equation, which will not affect the balance of the equation and can improve numerical performance in computations.

2.9 $\mathbb{R}$ is closed under the arithmetic operations

Our tests and discussions so far give the following remark:

> The set of real numbers $\mathbb{R}$ is **closed** under addition, subtraction, multiplication, and defined division.

Intuitively, the enclosure property of $\mathbb{R}$ holds because the bounds of real numbers are the infinities. Any newly produced number via an arithmetic operation on a pair of finite numbers can only stay within the bounds and hence enclosed.

2.10 Norm of a real number

Sometimes, we need to measure the *size* of a real number, regardless of its sign. This is done using the absolute value of the real number. It is defined as

$$abs(a) \equiv |a| \equiv \|a\| \equiv \begin{cases} a, & \text{if } a \text{ is positive} \\ -a, & \text{if } a \text{ is negative} \end{cases} \tag{2.1}$$

which can also be called a **norm** of the number known as the absolute norm. It is the length of the line segment measured from 0 to a (or $-a$) in $\mathbb{R}$. The same term "norm' will also be used for more complicated objects (functions, vectors, matrices, etc.).

Let us compute norms of some real numbers using the following Python code:

```
1  np.set_printoptions(precision=4, suppress=True)        # control digits
2
3  x1, x2 = -88.88, 98.98
4  print(f'The value of x1={x1}')
5  print(f'Absolute value of x1={np.abs(x1)}, Norm of x1={lg.norm(x1)}\n')
6  print(f'The value of x2={x2}')
7  print(f'Absolute value of x2={np.abs(x2)}, Norm of x1={lg.norm(x2)}')
```

```
The value of x1=-88.88
Absolute value of x1=88.88, Norm of x1=88.88

The value of x2=98.98
Absolute value of x2=98.98, Norm of x1=98.98
```

2.11 Real coordinate spaces

Real coordinate spaces can be used to manage and visualize real numbers graphically.

In physics, we may study number quantities that are purely *physical* and hence may not need to bring in coordinates. In dealing with many problems in science and engineering, the domain of the problem often has a finite dimension. One often needs a precise description of locations on devices or systems. We need to know not only the values of the numbers but also their locations and even their variation from one location to another. It is thus often useful to introduce **coordinates** with 0 at the origin. It serves as the reference point for the coordinate space. The most often-used coordinate system is the Cartesian coordinate system and the polar coordinate system. Here, we will brief on the **Cartesian coordinate** system.

2.11.1 *Numbers in one-dimensional real coordinate space $\mathbb{R}^1$*

Figure 2.2 shows some real numbers, as points, on a line in a 1D real coordinate space $\mathbb{R}^1$. Here, the superscript "1" is used to emphasize that it is now a 1D space. Members in $\mathbb{R}^1$ are the set of real numbers we discussed. Therefore, $\mathbb{R}^1$ is closed under the defined arithmetic operations, and it is also a **vector space** (of real numbers) over itself, implying that it is closed on linear combinations.

2.11.2 *Vectors in 2D real coordinate space $\mathbb{R}^2$*

Considering numbers that can vary freely independently, say ($x \in \mathbb{R}^1$, $y \in \mathbb{R}^1$). We can introduce two axes: x-axis usually placed horizontally and y-axis placed vertically. These two axes are in general kept **perpendicular**. They are also said to be **orthogonal** with each other. We now have a 2D real coordinate space $\mathbb{R}^2$, where coordinate points of (x, y) pairs can live.

To manage properly this kind of paired numbers, we need a "container" to hold them together. Such a container is called **vector**. It is defined as

$$\mathbf{r} = \begin{bmatrix} x \\ y \end{bmatrix} \tag{2.2}$$

It has two coordinate components: x and y. Therefore, a vector can uniquely present a coordinate point with two independent numbers in $\mathbb{R}^2$.

Alternatively, a vector can be represented geometrically. It has two characteristics: length and direction. Both are uniquely determined by the coordinates at its tip:

$$|\mathbf{r}| = \sqrt{x^2 + y^2}; \quad \alpha = \arccos\left(\frac{x}{\sqrt{x^2 + y^2}}\right) \tag{2.3}$$

where α is the angle between the vector and the x-axis.

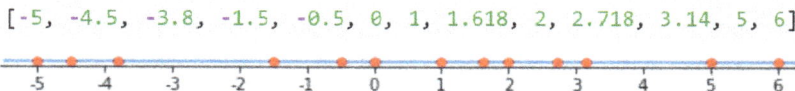

[-5, -4.5, -3.8, -1.5, -0.5, 0, 1, 1.618, 2, 2.718, 3.14, 5, 6]

Figure 2.2. Some real numbers shown as points on a line in the real coordinate space $\mathbb{R}^1$ with origin at 0.

Let us write a Python code to plot one point defined by paired numbers, and the vector formed $\mathbb{R}^2$:

```python
from matplotlib.patches import Arc, FancyArrowPatch
plt.rcParams.update({'font.size': 9})
plt.axes().set_aspect('equal')

x, y = 0.3, 0.2                       # a data-point in the x-y coordinate space
vector_r = [x, y]                     # a data-point as a vector in the 2D space
origin = [0, 0]                          # origin of the coordinate space

length_r = np.sqrt(x**2 + y**2)             # the length of the vector
alpha = np.arccos(x/length_r)*180/np.pi       # the angle of the vector
print(length_r, alpha)

plt.plot(origin[0], origin[1],'rx')
plt.plot(vector_r[0], vector_r[1],'ro')
plt.arrow(origin[0],origin[1],vector_r[0],vector_r[1], width=0.002,lw=2
            color='b', length_includes_head=True)

plt.title('A single point as a vector in 2D space')

plt.grid(color='r', linestyle=':', linewidth=0.5)    # grid for reference

#plt.hlines(0,-0.1,0.5, color='k')        # horizontal line for the x-axis
plt.arrow(-.1,0, .6,0, color='k', width=.002, lw=0,        # the x-axis
        length_includes_head=True)
plt.arrow(0,-0.1, 0,.4, color='k', width=.002, lw=0,        # the y-axis
        length_includes_head=True)
# Draw an angle arc:
arc = Arc((0, 0), width=x/2, height=y/1.5, angle=0,
        theta1=0, theta2=alpha, color='black')
plt.gca().add_patch(arc)
plt.text(x/2., y/1.6, r'$\mathbf{r}$', fontsize=13)
plt.text(x/3.5, y/10, 'α', fontsize=13)

# Arrowhead at the arc tip:
arrow = FancyArrowPatch((x/4.4, y/7), (x/5, y/4.9), arrowstyle='->',
                        mutation_scale=10, color='black')
plt.gca().add_patch(arrow)

plt.xlim(-0.1,0.5)                      # domain for display this data point
plt.ylim(-0.1,0.3)

plt.savefig('images/PointIn2D.png', dpi=500)
plt.show()                    . . .
```
0.36055512754639896 33.6900675259798

Note that a vector becomes now a mathematical object and a set of vectors can form a **vector space** just like a real number. We first define a zero vector that is a vector with all zero components. We need also **two rules for operations to vectors in a vector space**. The first rule is for "scalar

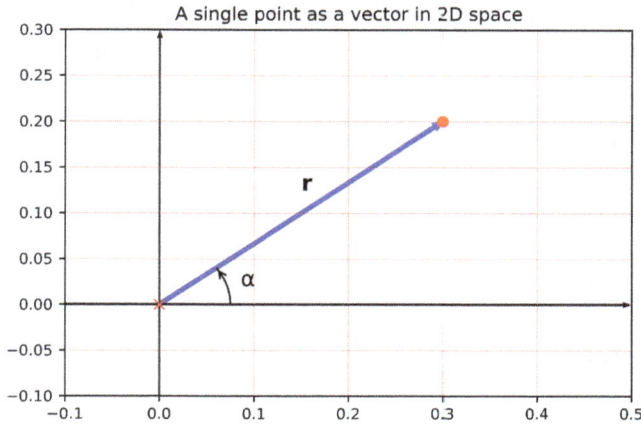

Figure 2.3. A point (red dot) defined by a pair of two real numbers $\begin{bmatrix} x \\ y \end{bmatrix} = \begin{bmatrix} 0.2 \\ 0.3 \end{bmatrix}$ in 2D coordinate space $\mathbb{R}^2$ with origin $\begin{bmatrix} 0 \\ 0 \end{bmatrix}$ marked by ×. It can also be uniquely represented by a vector (blue arrow) with length and direction.

over a field $\mathbb{F}$ multiplication to a vector": the scalar multiplies each of the components of the vector. The second rule is for "addition of two vectors": it is a component-wise addition, meaning that each of the components in these two vectors is added up correspondingly. With these operations defined for all the vectors, $\mathbb{R}^2$ also becomes a vector space (of vectors), implying that it is closed on linear combinations. This means that a linear combination of any vectors in the space will still be in the same space. This can easily be proven using the closure property of the numbers that form the vectors.

Note that the concept of vector is broad. In Fig. 2.3, the vector represents a coordinate point in $\mathbb{R}^2$, but a vector can be many other types of physical quantities, such as displacement, velocity, and acceleration. For example, when the horizontal coordinate is for a velocity component in the x-direction, and the vertical coordinate is for a velocity component in the y-direction, then the vector will be a velocity vector in an $\mathbb{R}^2$. It is then expressed as

$$\mathbf{v} = \begin{bmatrix} v_x \\ v_y \end{bmatrix} \tag{2.4}$$

Depending on the type of problem, we have all kinds of physical vectors. But in mathematics, we just call it a **vector**, a container that holds numbers in a defined order. It becomes a mathematically symbolized object, just like a symbolized real number (that can be mass, length, etc.). The only difference is that a vector has more components in it. In an $\mathbb{R}^2$ space, a vector has two components. Of course, these two rules for operations mentioned earlier apply.

We can plot many coordinate points live in the 2D coordinate space $\mathbb{R}^2$:

```
1  xL, xR = -10, 10
2  yL, xU = -10, 10
3  n_points = 20
4  x = np.arange(xL, xR)
5  y = np.arange(yL, xU)
6
7  plt.rcParams.update({'font.size': 9})
8  for i in range(n_points):
9      for j in range(n_points):
10         plt.plot(x[i], y[j],'ro')
11
12  plt.title('Real numbers as data points in a 2D rectangular domain')
13  plt.grid(color='r', linestyle=':', linewidth=0.5)
14  plt.savefig('images/PointsIn2D.png', dpi=500)
15  plt.show()
```

Figure 2.4 shows the arrays of coordinate points of real numbers in 2D coordinate space $\mathbb{R}^2$.

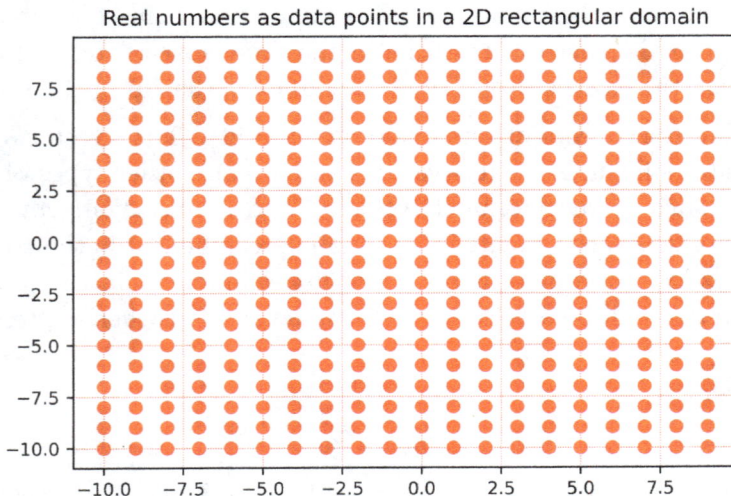

Figure 2.4. Arrays of coordinate points in 2D coordinate space $\mathbb{R}^2$. Each point can be represented by a coordinate vector.

The following code generates coordinate points randomly sampled over a rectangular domain and plotted in Fig. 2.5:

```
1  np.random.seed(8)
2  xL, xR = -10, 10
3  yB, yU = -10, 10
4  n_points = 200
5  samples=np.random.uniform(low=[xL,yB], high=[xR,yU], size=(n_points, 2)
6
7  plt.rcParams.update({'font.size': 9})
8  plt.scatter(samples[:,0],samples[:,1],s=8,c='m', marker='o', alpha=.8)
9
10 plt.title('Data points randomly sampled in a rectangular domain')
11 plt.grid(color='r', linestyle=':', linewidth=0.5)
12 plt.savefig('images/PointsIn2DRand.png', dpi=500)
13 plt.show()
```

Now, let us imagine that each of these coordinate points in Figs. 2.4 and 2.5 is represented by a vector as in Fig. 2.3. We see vectors covering the entire $\mathbb{R}^2$ space just like real numbers covering $\mathbb{R}^1$.

2.11.3 *Vectors in n-dimensional real coordinate space $\mathbb{R}^n$*

In general, we may have n-dimensional real coordinate space $\mathbb{R}^n$, where n is a natural number. When $n = 3$, it is the space we humans live in. For many real-life problems, we need to work with very high-dimensional spaces. The space for nodal displacements in an FEM model [3], for example, the dimension, can often be as high as millions and even billions.

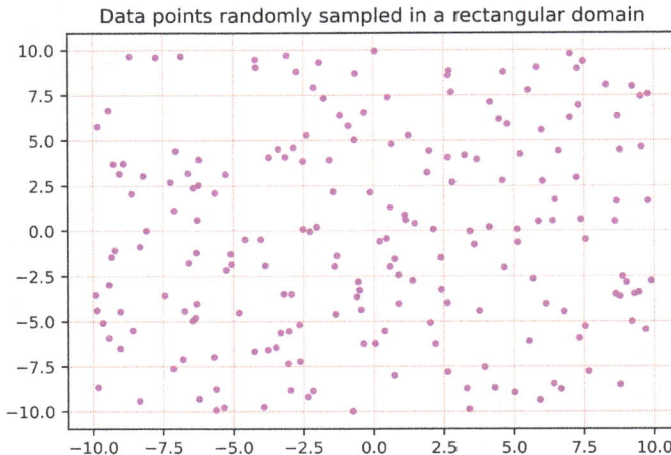

Figure 2.5. Randomly distributed coordinate points in 2D coordinate space $\mathbb{R}^2$.

The parameter space for a machine learning model [1] can be as high as billions. The GPT-4 model released in 2023 has 1.76 trillion parameters.

In an $\mathbb{R}^n$ space, a vector there shall have n components, each of which is a real number. It can be defined as

$$\mathbf{v} = \begin{bmatrix} v_1 \\ v_2 \\ \vdots \\ v_n \end{bmatrix} \tag{2.5}$$

where $v_i, i = 1, 2, \ldots, n$ are the components of the vector, each of which is a real (or complex, to be discussed in the following chapter) number. When a set of n-dimensional vectors with a zero vector forms a vector space, these two rules for operations mentioned earlier apply.

Again, the physical meaning of a vector shall depend on the physical problem.

We use $\mathbb{R}^n$ spaces frequently in computational methods for various types of problems, which will be discussed in detail in the future volumes. We can now imagine the complexity when these spaces are filled with numbers and vectors for us to operate on to find meaningful solutions in computational methods.

2.12 Remarks

1. Number is the most basic building block for all the mathematical objects.
2. The set of real numbers is closed under the four arithmetic operations. The outcome of arbitrary numbers undergoing these operations will still be a number in the same set. The numbers will not disappear subject to the machine's limits on bits.
3. Numbers form vectors in $\mathbb{R}^n$. The vectors become new mathematical objects. When these objects are operated upon, it eventually operates on numbers.

The concept of vector space is briefed for our use in this book. Former definitions, properties, and proofs are discussed in greater detail in a separate volume in this book series.

Chapter 3 discusses complex numbers.

References

[1]   Liu GR, *Machine Learning with Python: Theory and Applications*, World Scientific, 2023.

[2]   Liu GR, Tani J, Watanabe K *et al.*, A semi-exact method for the propagation of harmonic waves in anisotropic laminated bars of rectangular cross section, *Wave Motion*, 12(4), 361–371, 1990.

[3]   Liu GR, Tani J, Ohyoshi T *et al.*, Transient waves in anisotropic laminated plates, part 2: application, *Journal of Vibration and Acoustics*, 113(2), 235–239, 1991.

[4]   Liu GR, Lam KY and Ohyoshi T., A technique for analyzing elastodynamic responses of anisotropic laminated plates to line loads, *Composites Part B: Engineering*, 28(5), 667–677, 1997.

[5]   Liu GR and Siu Sin Quek, *The Finite Element Method: A Practical Course*, Butterworth-Heinemann, 2013.

Complex Numbers

Contents

Let us import the major necessary Python modules first for later use:

```
1  # Place curse in this cell, then Press Ctrl+Enter to import dependences.
2  import sys                          # For accessing the computer system
3  sys.path.append('../grbin/')     # Change to the directory in your system
4  from commonImports import *        # Import dependences from '../grbin/'
5  import grcodes as gr                # Import the module of the author
6  importlib.reload(gr)               # When grcodes is modified, reload it
```

The set of real numbers is closed under the arithmetic operations. However, it is not closed for nonlinear algebraic operations. The simplest example is that a square root of a negative real number is not in the set of real numbers. We need a bigger set of numbers to accommodate such operations. This leads to the topic of the set of complex numbers.

Complex numbers are extended from the real numbers discussed in detail in Chapter 2. The significance is that a complex number has an imaginary number with $i = \sqrt{-1}$, allowing all algebraic operations enclosed in the set of complex numbers. This enables convenient mathematical operations without worrying about the resulting numbers leaving the set:

1. Roots of all types of polynomials can be found.
2. Complex waveforms can be used for proper presentation of sinusoidal phenomena in the frequency domain, using techniques such as Fourier transform. The wave functions in quantum mechanics need to use complex numbers.
3. Convenient means can be devised to deal with differential equations.
4. In general, when analysis fails in using real numbers, complex numbers come to the rescue.

A complex number is another most basic element in computational methods. It is a building block for all the objects (functions, vectors, matrices, tensors, etc.). This chapter discusses the definition, properties, and behavior of complex numbers, especially their unique closure properties. It is written in reference to the documentation of Python, Numpy, Sympy, and other relevant wiki pages, including Complex_number and Linear space. "Discussions" with ChatGPT, Bing, and Bard together with the Google searched pages have also helped in the preparation of the chapter.

3.1 Definition

Complex numbers are created using real numbers. Therefore, all properties and behaviors of real numbers are used in deriving the properties of complex numbers. This is the **baseline** of our discussion on complex numbers.

A complex number c is defined as

$$c = a + ib \tag{3.1}$$

where a is its real part (Re) and b is its imaginary part (Im). Both a and b are real numbers in the set of real numbers $\mathbb{R}$. In Eq. (3.1), i is the newly introduced **imaginary unit**. It makes the complex number differ from the real number. It is defined as

$$i = \sqrt{-1} \quad \text{or} \quad i^2 = -1 \quad \text{or} \quad i \times i = -1 \tag{3.2}$$

Subject to Eq. (3.2), i is treated just as any of the real numbers in the arithmetic operations.

All the complex numbers formed using Eq. (3.1) are a set of complex numbers denoted as $\mathbb{C}$.

The complex numbers defined in Eq. (3.1) are no longer in $\mathbb{R}$ unless $b = 0$ because i (and its multiples with any real number, except 0) is not in $\mathbb{R}$. The operation defined in Eq. (3.2) is not one of the four arithmetic operations. Clearly, any real number can be treated as a complex number just as a special case when $b = 0$. Therefore, $\mathbb{R}$ is a subset of $\mathbb{C}$.

We will show later that $\mathbb{C}$ is a field because it satisfies the definition of a Field. Unlike real numbers, $\mathbb{C}$ is not an ordered field by definition because it is not always possible to compare the values of two arbitrarily given complex numbers, or the square of a complex number may be negative, as shown in Eq. (3.1).

As b in Eq. (3.1) is independent of a, the elements in $\mathbb{C}$ are essentially squared of that in $\mathbb{R}$. This is similar to the number of elements in $\mathbb{R}^2$. However, because a complex number is formed using Eq. (3.1), all the elements are still in 1D coordinate space $\mathbb{C}^1$, which is difficult to draw as we did in Fig. 2.1 for real numbers. We need a Re~Im plane to represent it geometrically.

3.2 Geometrical representation

3.2.1 *Re~Im plane*

A complex number can be represented **geometrically** in a Re~Im plane in a polar form. In this case, complex numbers have a standard basis of $[1, i]$, where 1 is on the Re-axis and i is on the imaginary axis, as shown in Fig. 3.1.

In Fig. 3.1, for any given a at Re-axis, b (real number) can change to one in between $-\infty$ and ∞. Each paired a and b produces only one complex number. Therefore, at any point in the Re-axis, there can be infinite number of complex numbers. This shows clearly that the number of complex numbers is "squared" of that of real numbers.

Using the polar form, a complex number can be graphically represented by an arrow in polar coordinates with $\|c\|$ as its length and φ as its polar angle, which is also called phase angle.

Using Fig. 3.1, a complex number can be expressed as follows based only on the geometry:

$$c = |c|(\cos \varphi + i \sin \varphi) \tag{3.3}$$

where $|c|$ is the absolute value of c. It is also called norm, modulus, or magnitude. Note that the sine and cosine are also defined purely on geometry (edge lengths and angles of triangles).

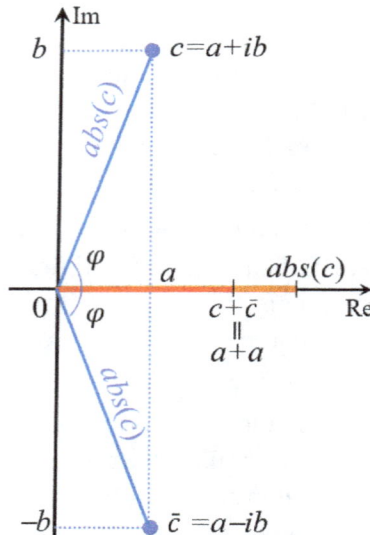

Figure 3.1. Geometric representation of complex numbers in the Re~Im plane. A complex number can be quantified by its absolute value $abs(c)$ and phase angle φ in the polar form.

3.2.2 *Periodicity of complex numbers*

Due to the geometric identity, the angle φ can be $(\varphi + 2k\pi)$, where k is an integer (both positive and negative). Thus, it is multi-valued in the Re~Im plane. This is also shown in Eq. (3.3) by noting the $(2k\pi)$ periodicity of sine and cosine.

When the phase angle φ is confined with the primary domain of $(-\pi, \pi]$, it can be calculated using

$$\varphi = \begin{cases} 2\arctan\left(\dfrac{b}{\sqrt{a^2 + b^2} + a}\right) & \text{if } a > 0 \quad \text{or} \quad b \neq 0 \\ \pi & \text{if } a < 0 \quad \text{and} \quad b = 0 \\ \text{undefined but finite} & \text{if } a = 0 \quad \text{and} \quad b = 0 \end{cases} \tag{3.4}$$

In the Python math module, it can be computed using atan2().

The complex plane shows that when a real number is subjected to operations other than these four defined arithmetic operations, it gets off the Re-axis and gets into the complex plane. Thus, real number set is not closed under such operations, but the complex set is.

3.2.3 *Euler's equation*

Euler's equation is frequently used in analysis of complex numbers (and functions). It has the following form:

$$e^{i\varphi} = \cos\varphi + i\sin\varphi \tag{3.5}$$

where e is Euler's number $e \approx 2.718$. The easiest proof of Eq. (3.5) needs to use the differentiable property of functions $e^{i\varphi}$, $\cos\varphi$, and $\sin\varphi$. We will do so in another volume after the derivatives of functions are discussed. For now, let us use Eq. (3.5) to examine the properties of complex numbers.

3.2.4 *Multi-values of numbers in Re~Im plane*

Due to the periodicity, complex numbers are multi-valued when viewed in the Re~Im plane. Using Eq. (3.5), we have

$$e^{i(\varphi + 2k\pi)} = \cos(\varphi + 2k\pi) + i\sin(\varphi + 2k\pi) \quad \forall \text{ integer } k \text{ and real } \varphi \tag{3.6}$$

Let $\varphi = 0$, we have $\cos(2k\pi) = 1$, and $\sin(2k\pi) = 0$, which leads to

$$1 = e^{i(2k\pi)} \quad \forall \text{ integer } k \tag{3.7}$$

We see now 1 is multi-valued, when viewed in the Re~Im plane. Similarly, let $\varphi = \pi$, we have $\cos(\pi + 2k\pi) = -1$, and $\sin(\pi + 2k\pi) = 0$, which leads to

$$-1 = e^{i(2k\pi+\pi)} \quad \forall \text{ integer } k \tag{3.8}$$

This shows -1 is also multi-valued.

Further, let $\varphi = \pi/2$, we have $\cos(\pi/2+2k\pi) = 0$, and $\sin(\pi/2+2k\pi) = 1$, which leads to

$$i = e^{i(2k\pi+\pi/2)} \quad \forall \text{ integer } k \tag{3.9}$$

The imaginary number i is also multi-valued. The same analysis is true for any real number viewed in Re~Im plane. As another interesting example, we have

$$i^i = e^{i(2k\pi+\pi/2)i} = e^{(-2k\pi-\pi/2)} \quad \forall \text{ integer } k \tag{3.10}$$

Considering only primary domain for the phase angle, we set $k = 0$. In this case, we have

$$i^i = e^{-\pi/2} \approx 0.2079 \tag{3.11}$$

which is a real number.

We know the squared imaginary number is a real number. It is interesting to find an imaginary number with an imaginary raised power is also a real number.

With the help of Eq. (3.5), the formula for a general real number can be expressed as

$$\begin{aligned} r &= |r|e^{i(2k\pi)} & \text{when } r > 0 \\ r &= |r|e^{i(2k\pi+\pi)} & \text{when } r < 0 \end{aligned} \tag{3.12}$$

We conclude that all numbers, real or complex, viewed in the Re~Im plane are multi-valued because of the periodicity of the sine and cosine functions. They have infinite number of possible values. These values are unique if the phase angle φ is confined within the primary domain of $(-\pi, \pi]$.

The multi-value property of complex numbers is important when examining the codomains of functions. It is also important in understanding multiplicity of solution to many problems.

3.3 Arithmetic operations for complex numbers

Based on the arithmetic operations for real numbers in $\mathbb{R}$, we *induce* the following arithmetic operations for all complex numbers in $\mathbb{C}$:

Complex zero:

By the definition of Eq. (3.1), a complex number is zero when both a and b are zero. Since a and b are all in $\mathbb{R}$, in which there is a zero, both a and b can be zero.

Scaling:

$$c_{\text{out}} = \alpha \times c = (\alpha \times a) + i(\alpha \times b); \quad \forall c \in \mathbb{C} \text{ and } \alpha \in \mathbb{R} \in \mathbb{C} \tag{3.13}$$

It is clear that c_{out} is in $\mathbb{C}$ because $(\alpha \times a)$ and $(\alpha \times b)$ are in $\mathbb{R}$ by the properties of real numbers and by the definition of Eq. (3.1).

Let's consider now two complex numbers, $c_1 = a_1 + ib_1$ and $c_2 = a_2 + ib_2$, both in $\mathbb{C}$.

Addition:

$$c_{\text{out}} = c_1 + c_2 = (a_1 + a_2) + i(b_1 + b_2) \tag{3.14}$$

Note c_{out} is also in $\mathbb{C}$ for the same reasons because $(a_1 + a_2)$ and $(b_1 + b_2)$ are in $\mathbb{R}$.

Subtraction:

$$c_{\text{out}} = c_1 - c_2 = (a_1 - a_2) + i(b_1 - b_2) \tag{3.15}$$

where c_{out} is also in $\mathbb{C}$ because $(a_1 - a_2)$ and $(b_1 - b_2)$ are in $\mathbb{R}$.

Multiplication:

$$c_{\text{out}} = c_1 \times c_2 = (a_1 + ib_1) \times (a_2 + ib_2) = (a_1 a_2 - b_1 b_2) + i(a_1 b_2 + a_2 b_1) \tag{3.16}$$

c_{out} is also in $\mathbb{C}$ because $(a_1 a_2 - b_1 b_2)$ and $(a_1 b_2 + a_2 b_1)$ are in $\mathbb{R}$.

Division:

$$c_{\text{out}} = \frac{c_1}{c_2} = \frac{a_1 + ib_1}{a_2 + ib_2} = \frac{(a_1 a_2 + b_1 b_2)}{(a_2^2 + b_2^2)} + i\frac{(-a_1 b_2 + b_1 a_2)}{(a_2^2 + b_2^2)}; \quad c_2 \neq 0 \tag{3.17}$$

c_{out} is also in $\mathbb{C}$ because $\frac{(a_1 a_2 + b_1 b_2)}{(a_2^2 + b_2^2)}$ and $\frac{(-a_1 b_2 + b_1 a_2)}{(a_2^2 + b_2^2)}$ are in $\mathbb{R}$.

It may be emphasized here that the division operation is not defined only when both $a_2 = 0$ and $b_2 = 0$. This property is useful when working around singularities. When the real number is 0, the division is not defined in $\mathbb{R}$, one can add in a small complex part b and perform the division in $\mathbb{C}$, as in Ref. [11].

3.4 Conjugate of a complex number

The conjugate of $c = a + ib$ is denoted as $\bar{c} = a - ib$. We have

$$\bar{c} + c = 2a \tag{3.18}$$

Also, the conjugate of a conjugated complex number gets back to itself:

$$\bar{\bar{c}} = c = a + ib \tag{3.19}$$

In addition, we also have

$$c\bar{c} = (a + ib)(a - ib) = (a - ib)(a + ib) = \bar{c}c = a^2 + b^2 \qquad (3.20)$$

It also produces a real number. It can be verified simply using the Numpy conj() function:

```
1  c1 = 8 + 18j
2  conj_c1 = np.conj(c1)
3  print ('complex number c1=', c1, ' Its conjugate=', conj_c1)
4
5  c2 = conj_c1
6  conj_c2 = np.conj(c2)
7  print ('complex number c2=', c2, ' Its conjugate=', conj_c2)
8  print (f"c1+conj_c1={c1+conj_c1},  a real number.")
9  print (f"c1*conj_c1={c1*conj_c1}, a real number.")
10 print (f"a^2 + b^2 = {c1.real**2+c1.imag**2},   a real number.")
```

```
complex number c1= (8+18j)  Its conjugate= (8-18j)
complex number c2= (8-18j)  Its conjugate= (8+18j)
c1+conj_c1=(16+0j),  a real number.
c1*conj_c1=(388+0j), a real number.
a^2 + b^2 = 388.0,   a real number.
```

3.5 Norm of a complex number

Finally, the norm of a complex number is the same as its absolute value and is defined as

$$\|c\| \equiv abs(c) \equiv |c| \equiv \sqrt{c\bar{c}} = \sqrt{a^2 + b^2} \qquad (3.21)$$

Both the sign and imaginary number are all removed in this norm measure, resulting in a non-negative real number. This offers a way to compare the "size" of complex numbers.

Let us demonstrate this:

```
1  print(f"Absolute value of  c1={np.abs(c1):.4f}",
2        f"    The norm of  c1={lg.norm(c1):.4f}")
3  print(f"Absolute value of -c1={np.abs(-c1):.4f}",
4        f"    The norm of -c1={lg.norm(-c1):.4f}")
```

```
Absolute value of  c1=19.6977    The norm of  c1=19.6977
Absolute value of -c1=19.6977    The norm of -c1=19.6977
```

3.6 Numerical test on the closure of complex numbers

We write the following Numpy code for numerical tests on the closure of complex numbers. These numbers may be generated randomly based on either the uniform or normal distribution.

```python
 1  def closure_testC(N, uniform=True):
 2
 3      '''Perform numerical test on the closure of complex numbers
 4         under arithmetic and conjugation operatiors.
 5         Complex numbers are randomly generated.
 6      Input:
 7         N: number of tests.
 8         uniform: if True, sample floats using uniform distribution.
 9                  otherwise, using the standard normal distribution.
10      Output: closed or not-closed
11      return: None
12      '''
13      np.random.seed(None)                          #  set seed off
14
15      fnt = np.isfinite
16
17      if uniform:
18          random_cmplx = np.random.random
19      else:
20          random_cmplx = np.random.normal
21
22      record = 0
23      for i in range(N+1):   #generates N+1 random complex numbers and test
24          a = random_cmplx() + 1j*random_cmplx()
25          b = random_cmplx() + 1j*random_cmplx()
26          apb = a + b
27          amb = a - b
28          atb = a * b
29          adb = a / b
30          c_a = np.conj(a)
31          if (not(fnt(apb) or fnt(amb) or fnt(atb) or fnt(adb) or fnt(c_a))):
32              record = i
33              print(f"Complex numbers is not closed!")
34
35      if (record == 0): print(f"Complex numbers are closed!")
```

Let us run the test 10 million times, using uniform sampling. This takes about 1–2 minutes to run on the author's laptop.

```python
1  N = 10_000_000
2  closure_testC(N, uniform=True)
```

Complex numbers are closed!

The result is true. Let us run the test again by sampling the floats using the standard normal distribution. This also takes about 1–4 minutes.

```
1  N = 10_000_000
2  closure_testC(N, uniform=False)
```

`Complex numbers are closed!`

The result is again true. Readers may increase the number of tests to gain some conviction or find otherwise. The running time should be roughly linear with N.

Note that if one is convinced that the set of real numbers is closed, there should be no doubt that the set of complex numbers is also closed because the complex numbers are defined using real numbers via the arithmetic operations, and the imaginary number i should not contribute much to the growth of the new numbers.

Similar to real numbers, normalization or scaling should be used in the computation if too big numbers are involved.

With these defined operations, a complex number shall have the similar properties of a real number. Although a complex number has an imaginary part, it is "real" in mathematics just as a real number. In fact, the real number can now be treated as a special case of a complex counterpart, and we have $\mathbb{R} \in \mathbb{C}$. When Python is used, one often needs only to define the data type as a complex number, and the follow-up computations are practically the same as for a real number. The computation with complex numbers is of course more resource and time-consuming.

3.7 $\mathbb{C}$ is closed under the arithmetic operations

Based on the discussions and tests, noting that any two complex numbers undergoing these operations still stay in $\mathbb{C}$, we have the following remark:

> The set of all complex numbers $\mathbb{C}$ is **closed** under addition, subtraction, multiplication and defined-division.

Python has all the major methods to perform the arithmetic operations for complex numbers. Let us write a code to demonstrate this. First, let's look at a symbolic computation using Sympy.

3.7.1 Sympy examples of arithmetic operations on complex numbers

```
1  a1, b1 = sp.symbols("a1, b1",real=True)
2  a2, b2 = sp.symbols("a2, b2",real=True)
3  c1 = a1 + sp.I*b1            # sp.I: imaginary number defined in Sympy
4  c2 = a2 + sp.I*b2
5  c1 + c2
```

$$a_1 + a_2 + ib_1 + ib_2$$

```
1  c1/c2
```

$$\frac{a_1 + ib_1}{a_2 + ib_2}$$

```
1  sp.re(c1/c2)                 # get the real part of a complex variable
```

$$\frac{a_1 a_2}{a_2^2 + b_2^2} + \frac{b_1 b_2}{a_2^2 + b_2^2}$$

```
1  sp.im(c1/c2)                 # get the imaginary part of a complex variable
```

$$-\frac{a_1 b_2}{a_2^2 + b_2^2} + \frac{a_2 b_1}{a_2^2 + b_2^2}$$

3.7.2 Numpy examples of arithmetic operations on complex numbers

Let us use the standard Python for numerical computations of complex numbers:

```
1  # Demonstrate in Python the standard operations on complex numbers
2  # complex(), real(), imag(), scaling, +, -, *, / and norm or abs().
3
4  a0, b0 = 0., 0.                            # Initializing real numbers
5  a1, b1 = 2., 3.
6  a2, b2 = 4., 5.
7  alpha = 8.                                         # scaling factor
8
9  c0 = complex(a0,b0)              # converting a and b into a complex number
10 c1 = complex(a1,b1)
11 c2 = complex(a2,b2)
12
13 # printing real and imaginary part of complex number
14 print("c0=",c0,"Its real part =",c0.real,"imaginary part =",c0.imag)
15 print("c1=",c1,"Its real part =",c1.real,"imaginary part =",c1.imag)
16 print("c2=",c2,"Its real part =",c2.real,"imaginary part =",c2.imag)
```

```
c0= 0j Its real part = 0.0 imaginary part = 0.0

c1= (2+3j) Its real part = 2.0 imaginary part = 3.0

c2= (4+5j) Its real part = 4.0 imaginary part = 5.0
```

```
1  print("Scaling factor = ", alpha, "\n")
2
3  print("Scaled c1 =",alpha*c1,"Its real part =",alpha*c1.real,
4        "imaginary part =",alpha*c1.imag,"\n")
5
6  print("c1 + c2 =", c1+c2,   "  Its real part =",
7        a1+a2, "and imaginary part =", b1+b2,"\n")
8
9  print("c1 - c2 =", c1-c2, "  Its real part =",
10       a1-a2, "and imaginary part =", b1-b2,"\n")
11
12 print("c1 * c2 =", c1*c2, " Its real part =",
13       a1*a2-b1*b2,", imaginary part =",a1*b2+b1*a2,"\n")
14
15 print(f"c1 / c2 = {c1/c2:.4f}")
16
17 print(f"Its real part = {(a1*a2+b1*b2)/(a2**2+b2**2):.4f},",
18       f"imaginary part = {(-a1*b2+b1*a2)/(a2**2+b2**2):.4f}")
```

```
Scaling factor =  8.0


Scaled c1 = (16+24j) Its real part = 16.0 imaginary part = 24.0

c1 + c2 = (6+8j)    Its real part = 6.0 and imaginary part = 8.0

c1 - c2 = (-2-2j)    Its real part = -2.0 and imaginary part = -2.0
```

```
c1 * c2 = (-7+22j)  Its real part = -7.0 , imaginary part = 22.0
c1 / c2 = 0.5610+0.0488j
Its real part = 0.5610, imaginary part = 0.0488
```

3.8 Examples of nonlinear operations

3.8.1 *The quadratic formula*

Readers may be familiar with the quadratic formula that gives the roots of a quadratic polynomial. It is written as

$$x = \frac{-b \pm \sqrt{b^2 - 4ac}}{2a} \tag{3.22}$$

You may have been taught before that $b^2 - 4ac \geq 0$ must be satisfied when the quadratic formula is used. Using complex numbers, the condition is no longer needed. In addition, all these constants a, b, and c are also complex numbers in $\mathbb{C}$. In other words, there is no restriction.

Let us use the following Python code to confirm this. First, examine the solution for $b^2 - 4ac \geq 0$ for Eq. (3.22):

```
1  a, b, c = 5., 0.,-20.                    # Initializing the constants
2  print(f"a={a:.4f}; b={c:.4f}; c={c:.4f}")
3  print(f"b^2-4ac = {b**2-4.*a*c}")
4
5  x1 = .5*(- b + np.sqrt(b**2-4.*a*c))/a
6  x2 = .5*(- b - np.sqrt(b**2-4.*a*c))/a
7  print(f"The solution: x1={x1:.4f}; x2={x2:.4f}")
```

```
a=5.0000; b=-20.0000; c=-20.0000
b^2-4ac = 400.0
The solution: x1=2.0000; x2=-2.0000
```

It is seen that we obtain two solutions that are all real as expected. The imaginary part becomes zero automatically. Let us now look at a case of $b^2 < 4ac$, in which the square root does not work for a negative argument in $\mathbb{R}$:

```
1  import cmath
2
3  a, b, c = 5., 10.,10.                        # Initializing the constants
4  print(f"a={a:.4f}; b={c:.4f}; c={c:.4f}")
5  print(f"b^2-4ac = {b**2-4.*a*c}")
6
7  x1 = .5*(- b + cmath.sqrt(b**2-4.*a*c))/a
8  x2 = .5*(- b - cmath.sqrt(b**2-4.*a*c))/a
9
10 print(f"The solution: x1={x1:.4f}; x2={x2:.4f}")
11 print(f"True for x1? {a*x1**2+b*x1+c:.4f}")
12 print(f"True for x2? {a*x2**2+b*x2+c:.4f}")
```

a=5.0000; b=10.0000; c=10.0000

b^2-4ac = -100.0

The solution: x1=-1.0000+1.0000j; x2=-1.0000-1.0000j

True for x1? 0.0000+0.0000j

True for x2? 0.0000+0.0000j

This time we obtain two solutions that are all complex numbers and both satisfy the original quadratic equation. Note that computation is in $\mathbb{C}$, and all these constants can be complex numbers:

```
1  ar, br, cr = 5., 10.,10.                     # The real part of the constants
2  ai, bi, ci = 2.,  1., 8.                      # The real part of the constants
3
4  # converting a and b into a complex number
5  a, b, c = complex(ar,ai), complex(br,bi), complex(cr,ci)
6  print(f"a={a:.4f}; b={c:.4f}; c={c:.4f}")
7
8  x1 = .5*(- b + cmath.sqrt(b**2-4.*a*c))/a
9  x2 = .5*(- b - cmath.sqrt(b**2-4.*a*c))/a
10
11 print(f"The solution: x1={x1:.4f}; x2={x2:.4f}")
12 print(f"True for x1? {a*x1**2+b*x1+c:.4f}")
13 print(f"True for x2? {a*x2**2+b*x2+c:.4f}")
```

a=5.0000+2.0000j; b=10.0000+8.0000j; c=10.0000+8.0000j

The solution: x1=-0.4582-1.0571j; x2=-1.3349+1.5743j

True for x1? 0.0000+0.0000j

True for x2? -0.0000-0.0000j

3.8.2 *Roots of a general polynomial*

In general, the same is true for computing the roots of polynomials of arbitrary order n for n roots. One would need to write a code to numerically find these roots if $n > 3$ because there is no simple formulation available.

Let us first look at a simple case using Sympy:

```
1  x = sp.symbols('x')
2  expr = x**2 + 4
3
4  roots = sp.solve(expr,x)
5  print(f"Roots of x^2 + 4 = 0 :\n {roots}")
```

```
Roots of x^2 + 4 = 0 :
 [-2*I, 2*I]
```

Use Numpy and find roots of a polynomial with the given coefficients numerically:

```
1  n = 2                           # quadratic with 3 constants/coefficients
2  # Enter the constants of the polynomial in an array form
3  consts = np.array([5, 8, 6])            # same constants as previous case
4  print(f"{n} roots obtained:\n {np.roots(consts)}")
```

```
2 roots obtained:
 [-0.8+0.74833148j -0.8-0.74833148j]
```

Let us try a case of fifth-order polynomial with randomly generated real coefficients:

```
1  np.set_printoptions(precision=3, suppress=True)      # control print outs
2  n = 5
3  consts = np.random.randn(n+1)                  # randomly generated constants
4  print(f"Coefficients of {n} polynomial:\n {consts}")
5  print(f"{n} roots obtained:\n {np.roots(consts)}")
```

```
Coefficients of 5 polynomial:
 [-0.488  0.333 -0.073  0.103  0.94  -0.316]
5 roots obtained:
 [-1.069+0.j     0.054+1.188j  0.054-1.188j  1.318+0.j     0.325+0.j    ]
```

Next, try a case of ninth-order polynomial with randomly generated complex coefficients:

```
1  np.set_printoptions(precision=3, suppress=True)     # control print outs
2  n = 9
3  consts=np.random.randn(n+1) + 1j*np.random.randn(n+1)   # random, complex
4  print(f"Coefficients of {n} polynomial:\n {consts}")
5  print(f"{n} roots obtained:\n {np.roots(consts)}")
```

```
Coefficients of 9 polynomial:
 [ 1.638+0.394j  0.381-1.729j  1.029-0.325j  1.063-1.813j  1.165-1.682j
  0.024+0.534j  0.74 -0.066j -0.682-0.041j  0.619+0.508j -1.479-0.834j]
9 roots obtained:
 [ 0.884+1.171j -0.502+1.16j  -0.822+0.461j -0.998-0.359j  0.168+0.914j
 -0.287-0.82j   0.223-1.046j  0.756+0.132j  0.598-0.562j]
```

As shown, computations with complex numbers are as easy as with real numbers in Python.

3.9 $\mathbb{C}$ is also algebraically closed

A set is said to be **algebraically closed** if every non-constant polynomial with coefficients in the set has a root in the set. Due to the well-known fact that any non-constant polynomial with coefficients in $\mathbb{C}$ always has roots in $\mathbb{C}$, we mention the following remark:

> The set of complex numbers is **algebraically closed**.

Note that finding roots of a polynomial is the same as solving the polynomial equation for variable x. It is essentially computing the inverse expression of the polynomial. Thus, the enclosure property of $\mathbb{C}$ ensures that the inverse of the polynomial can be found through operations in the same set.

Algebraically, closure is not true for the set of real numbers because it is easy to find polynomials with coefficients in $\mathbb{R}$ having roots not in $\mathbb{R}$, as shown in the examples earlier. In fact, as a simpler example, polynomial $x^2 + 4 = 0$ with all real coefficients, but has two roots of $x = \pm i2$ that are not in $\mathbb{R}$.

3.10 Complex number under nonlinear operations

A typical nonlinear operation to a number is with a raised power. The power can be real and also complex. The general formula can be as follows:

$$y = x^c \tag{3.23}$$

where c is in general in $\mathbb{C}$. The resulting y is also in general in $\mathbb{C}$. When $c = 2$, we have the square, and when $c = 1/2$, we have the square root as the simplest cases. When $c = 0$ or $c = 1$, we have the simplest linear cases.

Let us look at the outputs y for different powers c:

```python
c = np.array([2, np.e, 2j, 8.+2*1j])        # some numbers
x =  np.random.randn(1)[0]                   # a real number
for i in range(len(c)):
    print(f'({x:.4f})^({c[i]:.4f}) = {(x**c[i]):.4f}')
```

```
(0.9682)^(2.0000+0.0000j) = 0.9375+0.0000j
(0.9682)^(2.7183+0.0000j) = 0.9160+0.0000j
(0.9682)^(0.0000+2.0000j) = 0.9979-0.0645j
(0.9682)^(8.0000+2.0000j) = 0.7709-0.0498j
```

```python
c = np.array([2, np.e, 2j, 8.+2*1j])                  # some numbers
x =  np.random.randn(1)[0] + 1j*np.random.randn(1)[0] # a complex number
for i in range(len(c)):
    print(f'({x:.4f})^({c[i]:.4f}) = {(x**c[i]):.4f}')
```

```
(0.7154+0.0986j)^(2.0000+0.0000j) = 0.5021+0.1411j
(0.7154+0.0986j)^(2.7183+0.0000j) = 0.3846+0.1502j
(0.7154+0.0986j)^(0.0000+2.0000j) = 0.6049-0.4607j
(0.7154+0.0986j)^(8.0000+2.0000j) = 0.0508+0.0242j
```

It is seen that a nonlinear operation to an arbitrary number x produces in general a complex number.

3.11 Complex number from transcendental operations

A transcendental operation is a special type of nonlinear function frequently involving complex numbers. Here, we discuss the most often-used ones: exponential and trigonometric functions.

3.11.1 *Exponential operations*

In general, an exponential function is denoted as

$$f(x) = b^x \tag{3.24}$$

where b is called base, which is often a given constant. The raised power as the exponent x is a variable. In general, $x \in \mathbb{C}$. The base b is often confined in $\mathbb{R}$, but can also be in theory in $\mathbb{C}$.

Most widely used bases are 2, 10, and e, known as Euler's number $e \approx 2.718$. The most often used is called the natural exponential function, e^x.

Let us look at the outputs of some of the exponential operations:

```
1  b = np.array([2, 10, np.e, 2j, 8.+2*1j])        # some numbers
2  x =   np.random.randn(1)[0]                       # a real number
3
4  for i in range(len(b)):
5      print(f'({b[i]:.4f})^({x:.4f}) = {b[i]**x:.4f}')
```

```
(2.0000+0.0000j)^(-0.4761) = 0.7189+0.0000j
(10.0000+0.0000j)^(-0.4761) = 0.3341+0.0000j
(2.7183+0.0000j)^(-0.4761) = 0.6212+0.0000j
(0.0000+2.0000j)^(-0.4761) = 0.5271-0.4889j
(8.0000+2.0000j)^(-0.4761) = 0.3638-0.0426j
```

```
1  b = np.array([2, 10, np.e, 2j, 8.+2*1j])        # some numbers
2  x =   np.random.randn(1)[0] + 1j*np.random.randn(1)[0] # a complex number
3
4  for i in range(len(b)):
5      print(f'({b[i]:.4f})^({x:.4f}) = {b[i]**x:.4f}')
```

```
(2.0000+0.0000j)^(-1.4346-1.0022j) = 0.2842-0.2368j
(10.0000+0.0000j)^(-1.4346-1.0022j) = -0.0247-0.0272j
(2.7183+0.0000j)^(-1.4346-1.0022j) = 0.1283-0.2007j
(0.0000+2.0000j)^(-1.4346-1.0022j) = -1.7523-0.3433j
(8.0000+2.0000j)^(-1.4346-1.0022j) = -0.0483-0.0388j
```

It is seen that an exponent of various types of numbers to an arbitrary complex number x produces a complex number.

3.11.2 *Trigonometric functions*

The outputs of some trigonometric operations are as follows:

```
1  n = 2
2  x = np.random.randn(n+1) + 1j*np.random.randn(n+1)       # random, complex
3  x[0] = 0
4  x[1] = np.pi/4
5
6  for i in range(len(x)):
7      print(f'sin({x[i].real:.3f}+{x[i].imag:.3f}j) = {np.sin(x[i]):.4f}'
8      print(f'cos({x[i].real:.3f}+{x[i].imag:.3f}j) = {np.cos(x[i]):.4f}'
9      print(f'tan({x[i].real:.3f}+{x[i].imag:.3f}j) = {np.tan(x[i]):.4f}\n
```

```
sin(0.000+0.000j) = 0.0000+0.0000j
cos(0.000+0.000j) = 1.0000-0.0000j
tan(0.000+0.000j) = 0.0000+0.0000j

sin(0.785+0.000j) = 0.7071+0.0000j
cos(0.785+0.000j) = 0.7071-0.0000j
tan(0.785+0.000j) = 1.0000+0.0000j

sin(0.086+0.169j) = 0.0867+0.1697j
cos(0.086+0.169j) = 1.0107-0.0145j
tan(0.086+0.169j) = 0.0833+0.1691j
```

It is seen that the trigonometric operators on an arbitrary complex number x can produce complex numbers. Readers may try other trigonometric operators by changing the code cell given above a little.

Let us look at the roots for some transcendental operators:

```python
1  np.set_printoptions(precision=3)              # control print outs
2
3  x = sp.symbols('x')
4  n = 2
5  ci = np.random.randn(n+1) + 1j*np.random.rardn(n+1)    # random, complex
6  print(f"Coefficients of {n} polynomial:\n {ci}")
7
8  expr = ci[0]*sp.sin(x) - ci[1]*sp.sin(x)**2
9  roots = sp.solve(expr,x)
10 print(f"Roots a transcendental function:\n {roots}")
11
12 for i in range(len(roots)):
13     print(f'At {i}th root, expr = {expr.subs({x:roots[i]}).evalf()}')
```

```
Coefficients of 2 polynomial:
 [-0.241+1.037j  0.302-0.509j -1.36 +0.226j]
Roots a transcendental function:
 [0.0, 3.14159265358979, -1.20915805022861 + 1.21425910658376*I,
4.35075070381841 - 1.21425910658376*I]
At 0th root, expr = 0
At 1st root, expr = -2.95411240661038e-17 + 1.26997955005801e-16*I
At 2nd root, expr = 1.00564425688169e-15 + 4.99073096397963e-15*I
At 3rd root, expr = 8.8640252921582e-16 + 4.93716901267362e-15*I
```

3.12 $\mathbb{C}$ is closed under nonlinear and transcendental operations

Based on the discussion and examples give above, we have the following remark:

> The set of complex numbers $\mathbb{C}$ is **closed** under nonlinear operations including transcendental operations.

We discussed three enclosure properties of $\mathbb{C}$. This implies that the results of all the algebraic operations on numbers in $\mathbb{C}$ will stay in $\mathbb{C}$. **This is fundamentally important because it ensures that solutions exist for all the linear and nonlinear expressions including transcendental expressions discussed earlier**. These important properties can also be made use of while handling a number of problems through computational methods.

3.13 Some applications of the imaginary part

3.13.1 *Mathematical workaround*

From its definition in Eq. (3.2), the imaginary unit i appears very counterintuitive: it is a number that multiplies itself to become negative 1! In the real world, it cannot be observed. However, mathematically, it is a very useful number for at least working around. Solutions to many problems can be conveniently written using i, when the solution is expressed mathematically in some types of domains. For example, for wave propagation or vibration problems, the solution in the *frequency* domain can be expressed neatly with an exponential term of $e^{i\omega t}$ where ω is the angular frequency and t is the time. However, when the solution is finally transformed back to the real time domain (of course with a consistent and correct mathematical procedure), we obtain only the displacement responses that are in real numbers and physically measurable [8, 10, 11]. The imaginary parts automatically disappear. Therefore, the introduction of the imaginary unit i allows operations to be carried out in the complex numbers and hence offers a useful mathematical workaround, avoiding the breakdown in operation when confined only in the real numbers, in which the square root of a negative number is not allowed (defined).

Expanding the domain for mathematical operations is a useful trick. For example, when handling difficult problems in the singular value decomposition (SVD) of matrices, we expand the dimension. This converts the SVD problem to a simpler eigenvalue decomposition (EVD) of a symmetric matrix that can be solved using existing algorithms.

3.13.2 *Imaginary unit: A glue*

Special attention may be paid to the role of the imaginary number defined in Eq. (3.2). A complex number is made of two real numbers, but it can be treated just like a single number. This is because when i is squared during the multiplication and division operations, it becomes -1 and is back to being a real number. The same can happen in many other situations, such as the i^i example we studied earlier. Therefore, i plays a role of a kind of *glue* between real numbers and complex numbers. This is significantly different from a 2D vector that is also made of two real numbers.

3.13.3 *Singularity bypass*

In addition, a complex number can have an inverse even if its real part is zero. See the following:

```
1  c = complex (0, 0.2)
2  print('c=', c, '   1./c =', 1./c)
```

```
c= 0.2j    1./c = -5j
```

This is a useful feature that one can make use of in many numerical computations, when there is a need to avoid the singularities. One of the examples is the methods for analyzing elastodynamic or wave responses of anisotropic laminates [9, 11] and media [7, 11], where complex integration paths are designed to circumference the poles (singular points) that may be on the real integration paths. Another example is computing the inversed machine learning models that use the TrumpNets [4] or TubeNets [2, 3, 6]. For example, when $\tanh(z)$ is used in a TrumpNet as the activation function [5], it is defined in the entire real coordinate space $z \in (-\infty, \infty)$. However, its inverse is the $\text{arctanh}(z)$ function that is defined only for $z \in (-1, 1)$. When z is in the vicinity of -1 and $z \leq -1$ or near 1 and $1 \leq z$, which happens often in the inverse analysis using the so-called direct weights inversion (DWI) method [1–4], the computation of $\text{arctanh}(z)$ breaks down. In such cases, the argument z can be extended to the complex domain. This can effectively enable continuous computations avoiding breakdown.

3.14 Vectors in n-dimensional complex space $\mathbb{C}^n$

Similar to vectors in $\mathbb{R}^n$ discussed in Section 2.11, vectors can also be in $\mathbb{C}^n$. Each vector will have n components, each of which is a complex number.

Dealing with numbers and vectors in $\mathbb{C}^n$ is essentially the same as dealing with those in $\mathbb{R}^n$. Using the same two rules for linear operations discussed in Section 2.11, $\mathbb{C}^n$ becomes a vector space and closed under linear combinations.

3.15 Remarks

1. A complex number is the most fundamental building block for all the mathematical objects: functions, vectors, matrices, tensors, etc.
2. The set of complex numbers is closed under the four arithmetic operations, algebraical operations, and the nonlinear operations (including the transcendental ones). The outcome of arbitrary numbers undergoing all these operations and the corresponding inverse operations will still be numbers in the same set. The numbers will not disappear subject to the machine's limits of bits. This ensures the existence of a solution for any of the algebraic equations (linear or nonlinear), and it can be found through these algebraic operations if the equations are well posed and the physical problem does have a solution. How to find it and find it efficiently are the tasks of the computational methods.
3. Since $\mathbb{R}$ is a subspace of $\mathbb{C}$, if one has a difficulty in finding a solution in $\mathbb{R}$, it is possible that the solution is hidden in $\mathbb{C}$. Thus, it is often useful to conduct operations in $\mathbb{C}$ by defining related variables as complex data types. This can be easily done in Python.

References

[1]   Liu GR and Xi ZC, *Elastic Waves in Anisotropic Laminates*, 2001.
[2]   Liu GR and Lam KY, Two-dimensional time-harmonic elastodynamic Green's functions for anisotropic media, *International Journal of Engineering Science*, 34(11), 1327–1338, 1996.
[3]   Liu GR, Wang XJ and Xi ZC, Elastodynamic responses of an immersed to a Gaussian beam pressure, *Journal of Sound and Vibration*, 233(5), 813–833, 2000.
[4]   Liu GR, Lam KY and Tani J, An exact method for analysing elastodynamic response of anisotropic laminates to line loads, *Mechanics of Composite Materials and Structures an International Journal*, 2(3), 227–241, 1995.
[5]   Liu GR, Lam Khin Yong and Tani Junji, An exact method for analyzing elastic waves in anisotropic media excited by harmonic line loads, *JSME International Journal Series A Solid Mechanics and Material Engineering*, 40(3), 320–327, 1997.

[6] Liu GR, FEA-AI and AI-AI: Two-way deepnets for real-time computations for both forward and inverse mechanics problems, *International Journal of Computational Methods*, 16(8), 1950045, 2019.

[7] Liu GR, Duan SY, Zhang ZM, *et al.*, Tubenet: A special trumpetnet for explicit solutions to inverse problems, *International Journal of Computational Methods*, 18(1), 2050030, 2021.

[8] Duan S, Wang L, Wang F, *et al.*, A technique for inversely identifying joint stiffnesses of robot arms via two-way TubeNets, 2021.

[9] Duan S, Wang L, Liu GR, *et al.*, A technique for inversely identifying joint-stiffnesses of robot arms via two-way TubeNets, *Inverse Problems in Science & Engineering*, 2021.

[10] Liu, GR, *Machine Learning with Python: Theory and Applications*, World Scientific, 2023.

[11] Duan S, Shi L, Wang L, *et al.*, An uncertainty inversion technique using two-way neural network for parameter identification of robot arms, *Inverse Problems in Science and Engineering*, 29(13), 3279–3304, 2021.

Chapter 4

Elementary Functions

Contents

As usual, we will use Python in this study. The following Python modules are imported for later use:

```
1  import sys
2  sys.path.append('../grbin/')            # for the use of our own module
3  from commonImports import *                # often-used external modules
4  import grcodes as gr
5  importlib.reload(gr)              # reload when changes is made to grcodes
```

Once we have some basic concept for numbers, we can discuss about functions that play with numbers and produce new numbers. It is used to present the behavior of various problems in sciences and engineering.

The concept of functions was established as earlier as in the 17th century. The theory of functions have been studied intensively, and it is a rich topic. This chapter aims to introduce the basics of elementary functions that are frequently used in computational methods, with a focus on scalar functions, which take a number as independent variables and produce a single number.

We will first use a simple function to present a number of concepts for functions and then introduce and examine various types of elementary functions.

This chapter is written in reference to the documentation on related Python, Numpy, Sympy, mpmath, and other relevant wiki pages, including Functions, Trigonometric functions, Rational_function, Logarithm, and Function composition "Discussions" with the ChatGPT, Bing, and Bard together with the Google searched pages have also helped in the preparation of the chapter.

4.1 Definition of functions

Generally speaking, a scalar function takes a number input from a set of numbers called **domain** denoted as $\mathbb{X}$ and produces another number in another set called **codomain** denoted as $\mathbb{Y}$. It can be expressed as

$$f : \mathbb{X} \mapsto \mathbb{Y} \tag{4.1}$$

where we used map symbol $\mapsto$ because function can also be viewed as a type of mapping. In many literatures, we often use this following expression:

$$y \equiv f(x) \tag{4.2}$$

in which x is a number varying in domain $\mathbb{X}$ and y is a number belonging to codomain $\mathbb{Y}$. Thus, x is often called an **independent variable**. Since x sits in parentheses of $f(x)$, it is often called **argument** or **input** for $f()$. Number y is called a **dependent variable** for its dependence on the input x, and it is also called **output** of the function $f()$. Function $f()$ is also viewed as an operator. It operates on x and produces $y = f(x)$.

4.2 A simple example: A square function

A schematic diagram of a function is given in Fig. 4.1, which simply squares a given input. In this case, the function is expressed as

$$f(x) = x^2 \tag{4.3}$$

If our analysis is restricted in real numbers, its domain is confined in $\mathbb{R}$. The codomain of this function will be positive numbers in $\mathbb{R}$.

If our analysis is in complex numbers, both its domain and codomain can be $\mathbb{C}$.

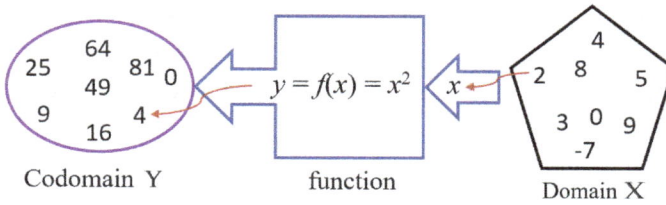

Figure 4.1. A simple function that squares an input x from domain $\mathbb{X}$ and produces an output y in codomain $\mathbb{Y}$.

A function can be viewed as a mathematical device, and it can come in different types to model the input–output relations in physical devices in science and engineering. Thus, there are many types of functions, and they can be generally very complicated. However, at an abstract higher level, a function can also be viewed as an object just as a number. Therefore, combinations of function objects can form new functions similar to combinations of numbers resulting in new numbers. The only difference is that the function object can produce many numbers when given multiple inputs. Having this concept in mind is helpful in understanding the essences of functions and their applications.

The following code creates first the simple function defined in Eq. (4.3) using Python and then computes the function values for given arguments (or inputs, or independent variables):

```
1  def function(x):           # define a code function of with argument x
2      y = x**2               # x: independent variable
3      return y               # y: output of function, dependent variable
4
5  x = np.linspace(-1, 1, 5)  # generate some values for array x
6  gr.printx('x')             # print array x
7  gr.printx('function(0)')   # compute & print the function value at x=0
8
9  ys = []                    # initialize a list for hosting f values
10 for i in range(len(x)):    # compute ys by passing x one-by-one
11     ys.append(function(x[i]))
12 gr.printx('ys')            # print function values for all x
13
14 ys = list(map(function,x)) # use map(fun, iterable) to compute y
15 gr.printx('ys')
```

```
x = array([-1. , -0.5,  0. ,  0.5,  1. ])
function(0) = 0
ys = [1.0, 0.25, 0.0, 0.25, 1.0]
ys = [1.0, 0.25, 0.0, 0.25, 1.0]
```

It is seen, using the code function defined, we can compute the function values by feeding the inputs one by one, or use the Python built-in method map() to map x to y in a single statement. In addition, the code function can be use simply as follows in Numpy:

```
1  ys = function(x)              # or simply use Numpy to generate an array
2  gr.printx('ys')
```

ys = array([1. , 0.25, 0. , 0.25, 1.])

We can also use the expression of the function directly, as we did in Chapter 1:

```
1  ys = x**2                     # or simply use the expression directly
2  gr.printx('ys')
```

ys = array([1. , 0.25, 0. , 0.25, 1.])

This allows us to see the relationship between the expression and the function as an operator. They are essentially the same. Note that functions can be very complicated need many code lines to define. Therefore, using expression directly is not always a viable option. Also, a function will usually be used for multiple times, and one may not want to remember and write the expression all the times. If the function is simple and is for single-time use, using the expression directly is a good practice.

Let us create a set of randomly generated complex values for the independent variable x and compute the square function values with those:

```
1  np.random.seed(8)                       # set seed to fix set of random numbers
2  n = 3
3  x = np.random.randn(n)+np.random.randn(n)*1j    # an array of n complex x
4  print(f'  x ={x}')
5  ys = function(x)
6  print(f'f(x)={ys}')
```

```
 x =[ 0.09120472-1.38634953j  1.09128273-2.29649157j
     -1.94697031+2.4098343j ]
f(x)=[-1.91364673-0.25288323j -4.08297555-5.0122432j
      -2.01660799-9.38375168j]
```

4.3 Distribution of functions

For square function given in Eq. (4.3), its domain $\mathbb{X}$ do not have to be discrete values. It can be the entire real space $\mathbb{R}$ and even the complex space $\mathbb{C}$. Consider a domain confined in the real space, its codomain will be real number in $(-\infty, 0]$. The function will be continuous, implying that it varies continuously with respect to continuously varying independent variable x. It can be plotted using the following code:

```
1  #help(ax.arrow)
```

```
1  fig, ax = plt.subplots(1,1,figsize=(3,2))
2  X = np.arange(-5, 5, .01)
3
4  ax.plot(X, function(X), label="$f(x)=x^2$")
5
6  ax.arrow(-5.,0., 10.,.0, color='k', width=.4,       # the x-axis (domain)
7           head_length=.2, length_includes_head=True)
8  ax.arrow(0.,0., -5.5,.0, color='k', width=.4,       # the x-axis (domain)
9           head_length=.2, length_includes_head=True)
10
11 ax.arrow(0,0.2, 0,25, color='g', width=.02,       # the y-axis (co-domain)
12          lw=2.5, length_includes_head=True)
13
14 ax.arrow(4,0.2, 0,15.8, color='r', ls='--', width=.03,    # x to curve
15          lw=.3, head_length=.6, length_includes_head=True)   # mapping
16 ax.arrow(4.,16., -4.,.0, color='r', ls='--', width=.1,     # curve to y
17          head_length=.15, lw=.3, length_includes_head=True)
18
19 ax.arrow(-4.5,0.2, 0,20., color='r',ls='--', width=.03,    # x to curve
20          lw=.3,  head_length=.6, length_includes_head=True)   # mapping
21 ax.arrow(-4.5,20.25, 4.5,.0, color='r', ls='--', width=.1,  # curve to y
22          head_length=.15,  lw=.3, length_includes_head=True)
23
24 ax.set_xlabel('x')
25 ax.set_title('Distribution of the square function')
26 ax.grid(color='r', linestyle=':', linewidth=0.3)
27 ax.legend(loc='upper right', bbox_to_anchor=(0.9, 0.95))
28 plt.savefig('images/f-x2.png', dpi=500)
29 plt.show()
```

Figure 4.2 plots the distribution of a simple square function in the form of a curve (blue). The corresponding domain and codomain for the function are also shown. This curve determines how the mapping is carried out in the x–y coordinate system.

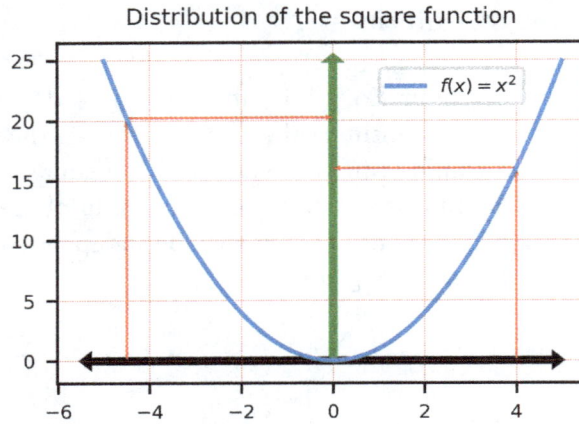

Figure 4.2.   A square function and its mapping process. Its domain is $(-\infty, -\infty)$ (black thick arrow) and plotted in $[-5, 5]$. Its codomain is $[0, \infty)$ (thick green arrow) and plotted in $[0, 25]$. The dashed red lines are mapping processes by the function (the blue curve).

4.4 Roots of a function

The roots of a function are the independent variable values, at which the function (expression) value is zero:

$$f(x) = 0 \qquad\qquad (4.4)$$

In general, a function can have multiple roots. These roots are the most important characteristics of a function.

For the square function given in Eq. (4.3), it has two roots at $x = 0$ (duplicated), at which $f(x) = x^2 = 0$. Sometimes, we say the function has a root multiplicity of 2.

4.5 Limits of a function

When studying a function, we often question about how it varies with respect to the independent variable. The limits of a function are such a property we want to know. This is because functions may not be bounded at some values of the independent variable. In other words, it does not have a finite **limit** when the variable approaches these values. For the square function, for example, its limits at $x = -\infty$ and ∞ are infinite. Sympy provides tools to examine the limits of a function:

```
1  from sympy import limit, oo, Symbol    # oo in sympy stands for infinity
2  sx = Symbol('sx')                       # sx: symbolic variable x
3
4  print(f'Limit at left: {limit(function(sx), sx,-oo)}')
5  print(f'Limit at right:{limit(function(sx), sx, oo)}')
```

```
Limit at left: oo
Limit at right:oo
```

4.6 Continuity of a function

Continuity of a function is another important property. Its former definition is as follows:

A function is said to be continuous at a point of interest $x = x_i$, if $f(x_i)$ is defined and the limits of the function at x_i from both sides of the function equal $f(x_i)$.

Sympy provides tool for checking the continuous domain of a function. Using the square function as an example, we can find

```
1  from sympy import S
2  from sympy.calculus.util import continuous_domain
3  continuous_domain(function(sx), sx, S.Reals)
```

$\mathbb{R}$

The square function is found continuous in the entire domain $\mathbb{R}$, as expected.

4.7 Reciprocal function

Since a number can have its reciprocal, a scalar function can also have a **reciprocal function**. It is expressed as follows:

$$[f(x)]^{-1} = \frac{1}{f(x)} \tag{4.5}$$

In this case, the domain for the reciprocal function may change because of the division operation. The codomain may also change accordingly. It is clear that the reciprocal function of a function is not defined at the roots of the function.

Let us write a f_reciprocal() function to compute the reciprocal function of f:

```python
np.seterr(divide='ignore')              # turn off the zero-division warning

def f_reciprocal(f, x):                 # define a reciprocal of function f
    y = 1/f(x)                          # use function to define function
    return y

x = np.arange(-2, 2, 1)
print(f'x = {x}')

# compute and print the function() created earlier for given x:
gr.printx('f_reciprocal(function, x)')
```

```
x = [-2 -1  0  1]
f_reciprocal(function, x) = array([0.25, 1.  ,  inf, 1.  ])
```

Readers may note the change of codomain. The reciprocal function of x^2 is not defined (bounded) at $x = 0$. We turned off the zero-division warning from Numpy just for this example, so that we can complete the computation without triggering an error. We now turn it back on:

```python
np.seterr(divide='warn')
#gr.printx('f_reciprocal(function, x)')    # try this with the waring on
```

```
{'divide': 'ignore', 'over': 'warn', 'under': 'ignore', 'invalid': 'warn'}
```

We can use Sympy to check the limit at $x = 0$:

```python
print(f'Limit at 0: {limit(f_reciprocal(function, sx), sx, 0)}')
```

```
Limit at 0: oo
```

```python
continuous_domain(f_reciprocal(function, sx), sx, S.Reals)
```

$$(-\infty, 0) \cup (0, \infty)$$

The reciprocal function of the square function is found continuous in domain $\mathbb{R}$, excluding $x = 0$.

4.8 Inverse function

As a function is essentially a mapping, we can naturally ask for the inverse mapping. This leads to a function called **inverse function**. It is expressed as follows:

$$x = f^{-1}(y) \tag{4.6}$$

Depending on the type of the function and the defined domains, such an inverse function may or may not exist. If it exists, the contents in the domain and codomain can change. For the function shown in Fig. 4.1, the inverse function is given in Fig. 4.3.

One may pay attention to the subtle differences in the position of the power of -1: for the reciprocal function, it is on the entire function value, while for the inverse function, it is only on the function (map) itself.

```
1  def f_inverse(y):                          # inverse of function x^2
2      x = np.sqrt(y)
3      return x
4
5  x = np.arange(-2, 2, 1)
6  print(f'x = {x}')
7  ys = function(x)                           # original function
8  print(f'f(x) = {ys}')
9
10 gr.printx('f_inverse(ys)')                 # compute and print
```

```
x = [-2 -1  0  1]
f(x) = [4 1 0 1]
f_inverse(ys) = array([2., 1., 0., 1.])
```

This gives back the original values of x. In this case, the inverse function and the original function are **mutually inverse**.

Let us check the continuous domain of the inverse function using Sympy:

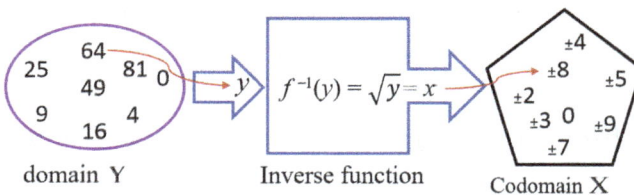

Figure 4.3. Inverse function of the square function. It takes an input y from domain $\mathbb{Y}$ that is the set of positive numbers, and produces x in codomain $\mathbb{X}$ that is a set of (positive and negative) numbers.

```
1  def sf_inverse(y):                              # inverse of function x^2
2      x = sp.sqrt(y)                              # square operator, nonlinear
3      return x
4
5  sy = Symbol('sy')                               # sy: symbolic y
6  from sympy.calculus.util import continuous_domain
7  continuous_domain(sf_inverse(sy), sy, S.Reals)
```

$[0, \infty)$

The inverse function of the square function is found continuous in domain $[0, \infty)$. Note that if we are working in complex domain, this inverse function is defined and continuous in $\mathbb{C}$ because the complex space is also closed under nonlinear operations, as discussed in Chapter 3.9.

```
1  np.set_printoptions(precision=4, suppress=True)        # print digits
2
3  np.random.seed(8)                    # set a seed to fix set of random numbers
4  n = 3
5  x = np.random.randn(n)+np.random.randn(n)*1j
6  print(f'  x =      {x}')
7
8  ys = function(x)                                       # original function
9  print(f'x^2 =      {ys}')
10
11 print(f'sqrt(x^2)= {np.sqrt(ys)}')
```

```
  x =       [ 0.0912-1.3863j  1.0913-2.2965j -1.947 +2.4098j]
x^2 =       [-1.9136-0.2529j -4.083 -5.0122j -2.0166-9.3838j]
sqrt(x^2)= [0.0912-1.3863j 1.0913-2.2965j 1.947 -2.4098j]
```

Inverse analysis is important in many studies, including inverse identification of parameters in mechanics systems, which finds an inverse of a very complicated function [3]. It was found that we often need to perform computations in complex domain for inverse analysis to avoid possible breakdowns in the process.

The rest of this chapter discusses the major types of elementary functions often encountered in computational methods.

4.9 Linear functions: The most widely used functions

4.9.1 *General form*

A linear function of x has the form

$$y \equiv f(x) := kx - b \tag{4.7}$$

where $x \in \mathbb{R}$ is the independent number variable, and $k \in \mathbb{R}$ and $b \in \mathbb{R}$ are given numbers. Function $f(x)$ or y is a number variable depending on x. Its domain $\mathbb{X}$ is $(-\infty, \infty)$ and codomain $\mathbb{Y}$ is $(-\infty, \infty)$ in $\mathbb{R}$.

In Eq. (4.7), we can treat number k and b as **parameters**. A function equipped with parameters has a more complicated behavior, when these parameters are tuned. Function y also becomes a linear function of x with two parameters of k and b. Since this function is computed using these four arithmetic operations on numbers, we have $y \in \mathbb{R}$. This is because the set of real numbers $\mathbb{R}$ is closed under the arithmetic operations.

If x or k or b is complex in $\mathbb{C}$, the linear function will also be complex in $\mathbb{C}$. This is because $\mathbb{R} \in \mathbb{C}$.

The following Python code computes and plots real linear function with varying x:

```python
1  fig, ax = plt.subplots(1, 2, figsize=(6,3))
2
3  def f(x):
4      return lambda k,b,: k*x-b              # f(x) with parameters k and b
5
6  x = np.arange(0, 10, .1)                         # domain of x
7  k0 = 1; k = np.arange(0, 5, 2)           # range for parameter k
8  b0 =20; b = np.arange(0, 5, 2)           # range for parameter b
9
10  for bi in b:
11      y = f(x)(k0, bi)
12      ax[0].plot(x,y, label="b="+str(bi))
13
14  ax[0].set_xlabel('x')
15  ax[0].set_ylabel("$y=f(x)$")
16  ax[0].set_title('$f(x)=kx-b$ with $k=$'+str(k0))
17  ax[0].grid(color='r', linestyle=':', linewidth=0.5)
18  ax[0].legend()
19
20  for ki in k:
21      y = f(x)(ki, b0)
22      ax[1].plot(x,y, label="k="+str(ki))
23
24  ax[1].set_xlabel('x')
25  ax[1].set_title('$f(x)=kx-b$ with $b=$'+str(b0))
26  ax[1].grid(color='r', linestyle=':', linewidth=0.5)
27  ax[1].legend()
28  plt.savefig('images/f_kx-b.png', dpi=500)
29  plt.show()
```

The major features of a linear function is as follows:

1. It is a straight line.
2. Its slope is constant.
3. It varies monotonically.

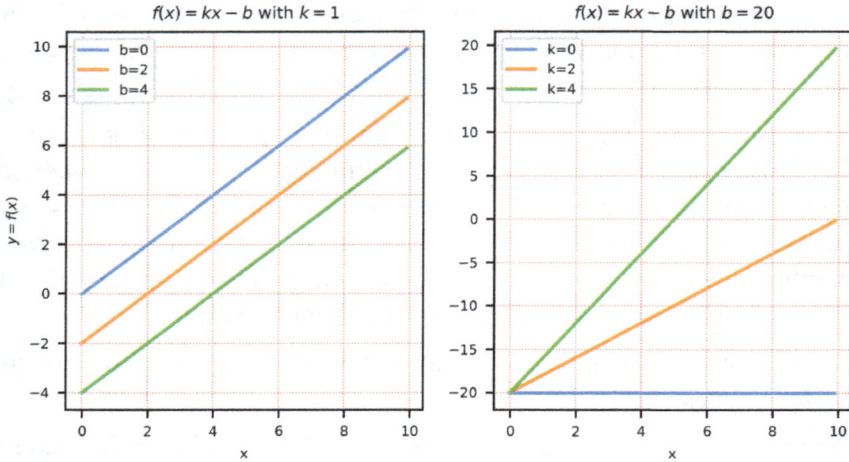

Figure 4.4.   Linear functions are straight lines in $x - y$ plot. Parameters k is the slope of the lines, and b gives an intersection on the y-axis.

4.9.2 *Some applications*

Linear functions defined in Eq. (4.7) are the most widely used functions in various applications in science and engineering. For example, these functions are used in numerical methods for modeling and simulation, such as in finite element methods (FEM) [5], smoothed finite element methods (S-FEM) [4], and meshfree methods [1, 7] for structure analysis. In such methods, the so-called shape functions are used, which have the same form of Eq. (4.7). In such applications, $f(x)$ will be a function for the displacements, and the parameters k and b will be the constants determined using conditions for determining the shape functions.

In establishing machine learning (ML) models [2], Eq. (4.7) is used in the affine transformation, the most fundamental operation in artificial intelligence (AI) algorithms.

4.9.3 *Root of the linear function*

The root of the linear function can be found easily by setting $f(x) = 0$:

$$kx - b = 0 \qquad (4.8)$$

For given parameters k and b, we can solve for x via the arithmetic operations, which gives

$$x = \frac{b}{k} \qquad (4.9)$$

This solution for the root exists as long as $k \neq 0$, where the division is defined. If $k = 0$, $f(x) = b$. It will never be zero for any non-zero b. Further, if $b = 0$, $f(x) = 0$. The function is trivial and no point for any further study.

Note that for $k \neq 0$, the linear function is bounded (its norm is smaller than ∞) if and only if the independent variable x is bounded.

4.9.4 *Reciprocal of linear function*

Reciprocal of a linear function becomes

$$[f(x)]^{-1} = \frac{1}{f(x)} = \frac{1}{kx - b} \tag{4.10}$$

This is a rational function (see definition later), and hence, its complexity greatly increased. First, this function is not defined (bounded) at the root of function $f(x)$, where $kx - b = 0$. Second, it varies with x in a nonlinear way. It is the best to view this using the following code to plot some such functions:

```
1  fig, ax = plt.subplots(1, 1, figsize=(6,3))        # a new plot
2
3  def f_recprcl(x):              # reciprocal func. of linear functions
4      return lambda k,b,: 1/(k*x-b)              # with parameters k & b
5
6  x = np.arange(-1, 5, .01)                            # domain of x
7  k0 = 1; k = np.arange(0, 5, 2)               # range for parameter k
8  b0 =20; b = np.arange(0, 5, 2)               # range for parameter b
9
10 for bi in b:
11     y = f_recprcl(x)(k0, bi)
12     ax.plot(x,y, label="b="+str(bi))
13
14 ax.scatter([0,2,4], [0,0,0], c = 'r', label='Roots')
15 ax.set_xlabel('x')
16 ax.set_ylabel("$y=[f(x)]^{-1}$")
17 ax.set_title(r'$[f(x)]^{-1} = \frac{1}{kx-b}$  with $k=$'+str(k0))
18 ax.grid(color='r', linestyle=':', linewidth=0.5)
19 ax.set_ylim(-10, 10)
20 ax.legend()
21
22 plt.savefig('images/r_f_kx-b.png', dpi=500)
23 plt.show()
```

It is shown clearly in Fig. 4.5 that these functions are not defined, respectively, at $x = 0$, $x = 2$, and $x = 4$, known as singular points. The functions decay drastically when away from the singular points, which is typical for a rational function. This feature is particularity useful in modeling infinite domain for bounded solutions, such as in creating the infinite elements in FEM [5].

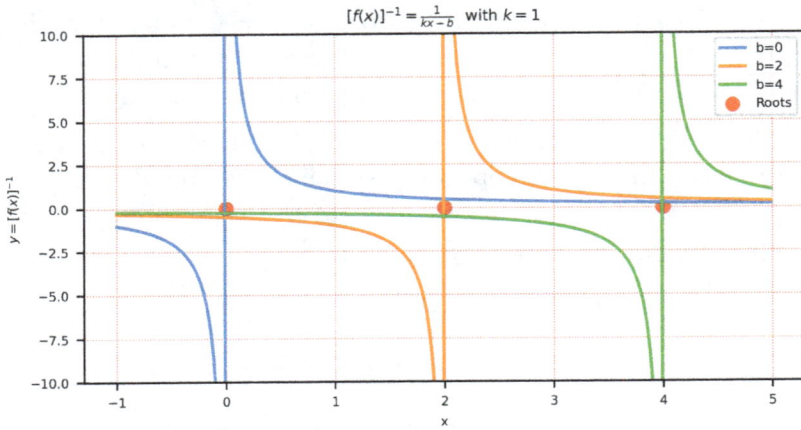

Figure 4.5.   Reciprocal functions of three linear functions.

Let us check the limits of the reciprocal function at a singular point at $x = 2$:

```
1  from sympy import limit, oo, Symbol
2  sx = Symbol('sx')                              # sx: symbolic variable x
3  print(f'Limit at x=2 (k=1,b=2): {limit(f_recprcl(sx)(1,2), sx, 2)}')
```

Limit at x=2 (k=1,b=2): oo

The following code checks the continuous domain of this reciprocal function with the singular point at $x = 2$:

```
1  from sympy.calculus.util import continuous_domain
2  continuous_domain(f_recprcl(sx)(1,2), sx, S.Reals)
```

$(-\infty, 2) \cup (2, \infty)$

This reciprocal function is found continuous in domain $\mathbb{R}$ excluding $x = 2$, as expected from the limit analysis.

4.9.5   *Inverse of linear function*

The inverse of a linear function becomes

$$x = \frac{y + b}{k} = \frac{1}{k}y + \frac{b}{k} \tag{4.11}$$

It is not defined (bounded) at $k = 0$ because y is not related to x at the first place. For non-zero k, the inverse function is also a linear function. Let us use the following code to plot some such functions to view the details.

```python
fig, ax = plt.subplots(1, 1, figsize=(4,3))              # a new plot

def f_inverse(y):                    # inverse functions of linear functions
    return lambda k,b,: (y+b)/k                  # with parameters k & b

x = np.arange(-1, 5, .01)                            # domain of x
k0 = 1; k = np.arange(0, 5, 2)               # range for parameter k
b0 =20; b = np.arange(0, 5, 2)               # range for parameter b

for bi in b:
    y = f_inverse(x)(k0, bi)
    ax.plot(x,y, label="b="+str(bi))

ax.set_xlabel('x')
ax.set_ylabel("$x=f^{-1}(y)$")
ax.set_title(r'Inverse of linear functions: '
             r'$f^{-1}(y) = \frac{y+b}{k}$  with $k=$'+str(k0))
ax.grid(color='r', linestyle=':', linewidth=0.5)
ax.legend()

plt.savefig('images/i_f_kx-b.png', dpi=500)
plt.show()
```

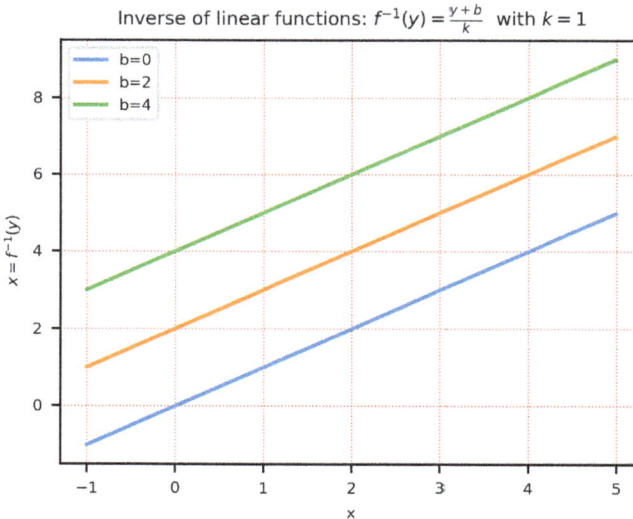

Figure 4.6. Inverse functions of three linear functions.

It is shown clearly that the inverse functions are still linear, as expected. Its slope is $1/k$, and inverse function exists if and only if k is non-zero. This finding is simple but fundamentally important for inverse problems. When complicated inverse problems are locally linear, this linearity preservation with inversed slope (gradient) is valuable [3].

4.10 Monomial functions

4.10.1 *Definition*

Another widely used elementary function is known as **monomial** function. It has the simple form of

$$f(x) = x^n \tag{4.12}$$

where n is a finite integer in a set of $\{0, 1, 2, \ldots\}$ (the natural numbers). Note that x^n with different n are linearly independent of each other, implying that any x^n cannot be expressed in a linear combination of the rest with different n. Therefore, they are used as the **basis function** to form a polynomial function (that will be discussed in detail later) via a linear combination of some of them. Monomial x^n is the nth-order basis, and 1 is the zeroth-order basis.

These monomials have roots at $x = 0$ when $n > 0$.

4.10.2 *Reciprocal of monomial functions*

Note that when n is negative, we have the **reciprocal** of monomial functions. In this case, these functions are singular at $x = 0$.

The entire family of monomial functions with $n \in \{\ldots, -2, -1, 0, 1, 2, \ldots\}$ is qualified as the basis for function approximation.

It is seen that when the power is negative, such as $\frac{1}{x}$ and $\frac{1}{x^2}$, it is also called rational function. It is not defined (or unbounded) at $x = 0$, which is the root of the monomial with positive integer power. They are continuous only in $(-\infty, 0)$ and $(0, \infty)$. The function is the reciprocal function of its counterpart positive n. In fact, monomials with powers of opposite signs are **mutually reciprocal**.

Let us plot a few monomial functions.

```python
1  Label=[r'$1$',r'$x$',r'$x^2$',r'$\frac{1}{x^2}$',r'$\frac{1}{x}$']
2  colors = ['b', 'r', 'g', 'c', 'm']
3
4  n   = 2; x_start = 0.1; x_end = 2.0; N = 100
5  x   = np.linspace(-x_end,x_end, 2*N)              # x on both sides
6  xl  = np.linspace(-x_end,-x_start,N)              # x on the left (negative)
7  xr  = np.linspace( x_start, x_end, N)             # x on the right (positive)
8
9  fig, ax = plt.subplots(1,1,figsize=(4,2.8))
10 ax.grid(c='r', linestyle=':', linewidth=0.5)
11
12 for k in range(0,n+1):                            # plot for n = 0, 1, 2
13     ax.plot(x, [xi**k for xi in x], c=colors[k], label = Label[k])
14
15 for k in range(-n,0):                             # plot for n = -2, -1
16     ax.plot(xl, [xi**k for xi in xl], c=colors[k], label = Label[k])
17     ax.plot(xr, [xi**k for xi in xr], c=colors[k])
18
19 ax.set_xlabel('x')
20 ax.set_ylabel(r"$1/f(x)$"); ax.set_ylim(-4, 6)
21 ax.axvline(x=0, c="k", lw=0.6); ax.axhline(y=0, c="k", lw=0.6)
22 ax.legend()
23
24 plt.savefig('images/monomials.png', dpi=500)
25 plt.show();
```

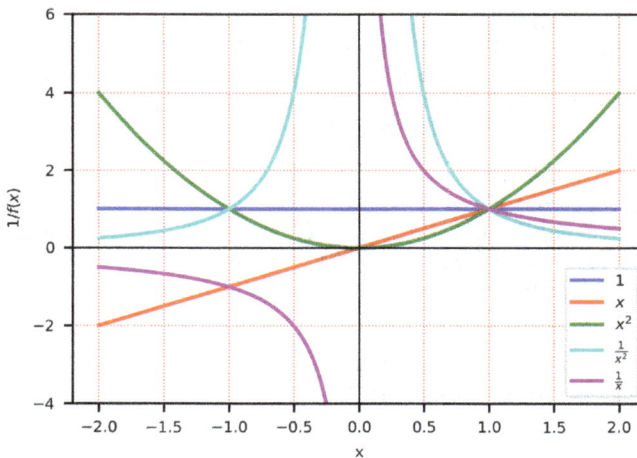

Figure 4.7. Examples of monomial functions. When the power is negative, such as $\frac{1}{x}$ and $\frac{1}{x^2}$, the function is the reciprocal function of its positive n counterpart. It is not defined (or unbounded) at $x = 0$, and they are continuous only in $(-\infty, 0)$ and $(0, \infty)$.

4.10.3 *Inverse of monomial functions*

Inverse of a monomial function can be expressed as follows:

$$x = \sqrt[n]{y} = y^{\frac{1}{n}} \tag{4.13}$$

where $n \in \{\ldots, -2, -1, 1, 2, \ldots\}$.

Let us plot inverse functions of a few monomial functions:

```python
1  Label=['0',r'$y$',r'$y^{½}$',r'$y^{-½}$',r'$\frac{1}{y}$']
2  colors = ['b', 'r', 'g', 'c', 'm']
3  l_style_n = [':', '-',':', '-']
4
5
6  n  = 2; y_start = 0.02; y_end = 2.0; N = 100
7  y  = np.linspace(-y_end, y_end, 2*N, dtype=complex)    # y on both sides
8  yl = np.linspace(-y_end, -y_start,N, dtype=complex)      # y on the left
9  yr = np.linspace( y_start, y_end, N)            # y on the right (positive)
10
11 fig, ax = plt.subplots(1,1,figsize=(4,3))
12 ax.grid(c='r', linestyle=':', linewidth=0.5)
13
14 def f_y(y):
15     '''Pick up nonzero values of a complex variable y.'''
16     if   np.isclose(y.imag, 0): y_value = y.real
17     elif np.isclose(y.real, 0): y_value = y.imag
18     return y_value
19
20 for k in range(1,n+1):                               # plot for n = 1, 2
21     ax.plot(yl.real,[f_y(yi**(1/k)) for yi in yl],
22             c=colors[k], linestyle=l_style_n[k], label=Label[k])
23     ax.plot(yr.real,[f_y(yi**(1/k)) for yi in yr],
24             c=colors[k], linestyle='-')
25
26 for k in range(-n,0):                                # plot for n =-2,-1
27     ax.plot(yl.real,[f_y(yi**(1/k)) for yi in yl],
28             c=colors[k], linestyle=l_style_n[k], label=Label[k])
29     ax.plot(yr.real,[f_y(yi**(1/k)) for yi in yr],
30             c=colors[k], linestyle='-')
31
32 ax.set_xlabel('y')
33 ax.set_ylabel(r"$f^{-1}(y)$"); ax.set_ylim(-6, 6)
34 ax.axvline(x=0, c="k", lw=0.6); ax.axhline(y=0, c="k", lw=0.6)
35 ax.legend()
36
37 plt.savefig('images/i_monomials.png', dpi=500)
38 plt.show();
```

Note that the power to the monomial (forward, reciprocal, and inverse) functions can be extended to complex numbers. In such cases, the function may be complex-valued. The power can even be complex. In this case, the function is complex-valued, and both the domain and codomain are $\mathbb{C}$ because complex domain is closed under nonlinear operations.

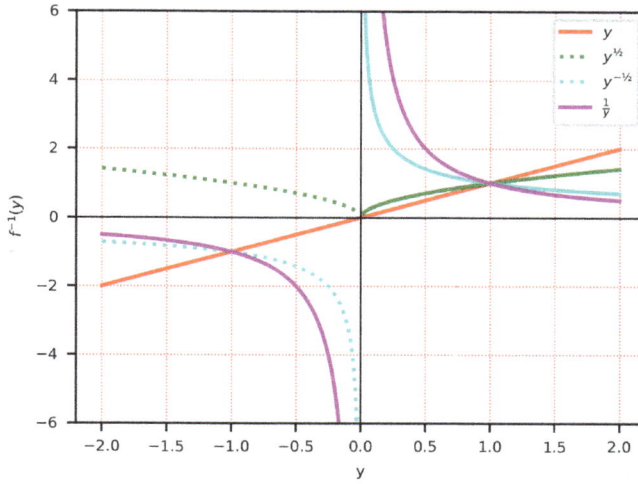

Figure 4.8. Examples of inverse functions of a few monomial functions. For negative y values, the inverse function produces imaginary number (dotted lines) when n is an even integer. When n is negative, the inverse function is not defined (or unbounded) at $x = 0$, and they are continuous only in $(-\infty, 0)$ and $(0, \infty)$.

Following are some examples involving complex powers:

```
1  x  = -4.
2  power_r = 0.5;  power_c = 0.8+0.5j
3  print(f'Monomial function with real power    = {x**power_r:.4f}')
4  print(f'Monomial function with complex power ={x**power_c:.4f}')
```

```
Monomial function with real power    = 0.0000+2.0000j
Monomial function with complex power =-0.6288-0.0408j
```

The major features of a monomial (forward, reciprocal, and inverse) function are as follows:

1. It varies **monotonically** in either $(-\infty, 0)$ and $(0, \infty)$.
2. When n is an even integer, it is an **even function**, implying that $f(x) = f(-x)$.
3. When n is an odd integer, it is an **odd function**, implying that $f(x) = -f(-x)$.
4. All x^n with different n are linearly independent and qualified as a **basis** for constructing new function.
5. Its reciprocals and inverses enrich this family of functions further.

In conclusion, the family of monomial functions are very versatile and hence frequently used as parts of many complicated forms of functions, such as polynomials and types of composite functions.

4.11 Complex-valued functions, closure in $\mathbb{C}$

4.11.1 *A typical example*

In studying the inverse functions of the square or in general monomial functions, we found the following form of functions:

$$y = \sqrt[n]{x} \tag{4.14}$$

where n is an integer. We note the following:

If the domain is all positive real numbers $[0, \infty)$, the codomain is $\mathbb{R}$. This means that the function is real-valued. If, however, the domain has negative real numbers in $(-\infty, 0)$, the codomain can be in $\mathbb{C}$. This means that the function can be complex-valued. Figure 4.9 shows a typical case.

4.11.2 *Closure of the square function in $\mathbb{C}$*

When the domain for the square function is complex domain $\mathbb{C}$, its codomain is also $\mathbb{C}$. It is closed in $\mathbb{C}$. It has no singular point.

The following code plots the distribution of the square function in a complex domain:

```python
xR = np.linspace(-np.pi, np.pi, 200)    # range in real axis
xI = np.linspace(-np.pi, np.pi, 200)    # range in imaginary axis
lxR, lxI = len(xR), len(xI)

XR, XI = np.meshgrid(xR, xI)
z2 = (XR + XI*1j)**2
Z = (     z2.real,  z2.imag, np.abs(z2), np.angle(z2))
z_labels=['Re(z^2)','Im(z^2)','abs(z^2)','angle(z^2)']

plt.rcParams.update({'font.size': 6})
plt.rcParams["figure.autolayout"] = True
fig_s = plt.figure(figsize=(6,4))

for i in range(len(Z)):
    ax = fig_s.add_subplot(2,2,i+1, projection='3d')
    ax.set_xlabel('Re(z)', labelpad=-11)
    ax.set_ylabel('Im(z)', labelpad=-11)
    ax.tick_params(axis='x', pad=-5)
    ax.tick_params(axis='y', pad=-5)
    ax.tick_params(axis='z', pad=-5)
    #ax.set_zlabel(z_labels[i], rotation=0)
    plt.title(z_labels[i], pad=-25)

    ax.plot_surface(XR,XI,Z[i],color='b',rstride=1,cstride=1,
                    shade=False,cmap="jet", linewidth=1)

fig_s.tight_layout()
plt.savefig('images/z2_complex.png',dpi=500,bbox_inches='tight')
plt.show()
```

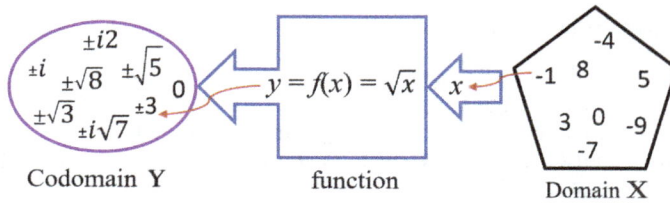

Figure 4.9. A complex-valued function as the inverse of the square function.

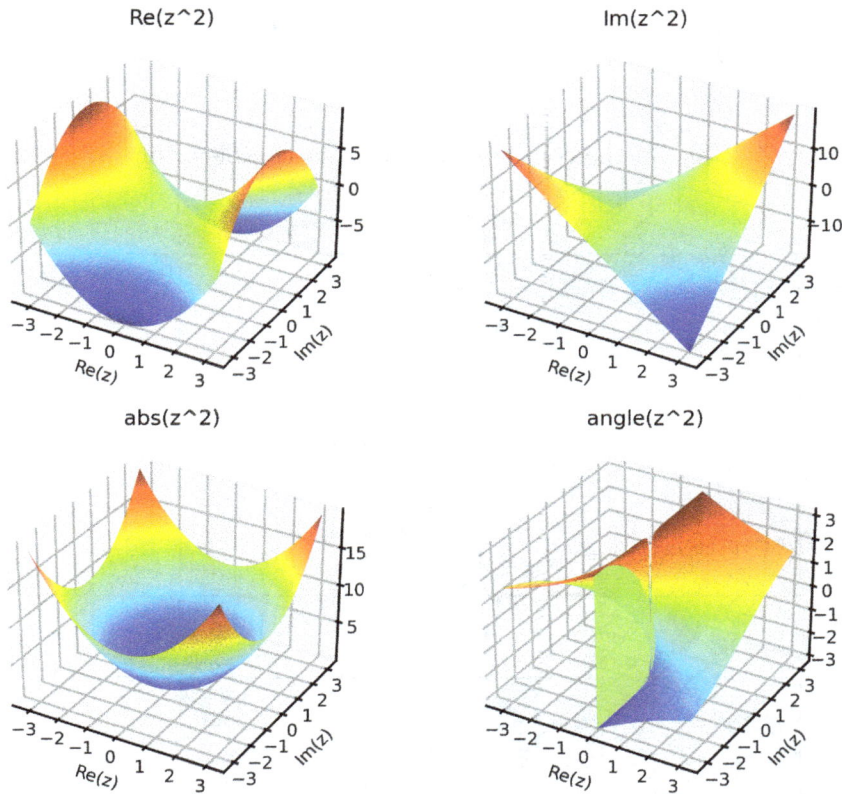

Figure 4.10. Square function plotted in complex domain.

Figure 4.10 plots the surface of z^2 in complex domain. Both the argument and function value are complex numbers.

It found that for any $z \in \mathbb{C}$, the output of z^2 is always in $\mathbb{C}$.

An alternative view is contour plots. The following code does this:

```
1  fig_s = plt.figure(figsize=(6,6))
2
3  for i in range(len(Z)):
4      ax = fig_s.add_subplot(2,2,i+1)
5      ax.set_xlabel('Re(z)'); ax.set_ylabel('Im(z)')
6      ax.axvline(x=0,c="k",lw=.3); ax.axhline(y=0,c="k",lw =.3)
7      plt.title(z_labels[i])
8
9      ax.contour(XR,XI,Z[i], levels=40, linewidths=0.8)
10
11 fig_s.tight_layout()
12 plt.savefig('images/z2_complexC.png',dpi=500,bbox_inches='tight')
13 plt.show()
```

Figure 4.11 shows the contour plots of z^2 in complex domain.

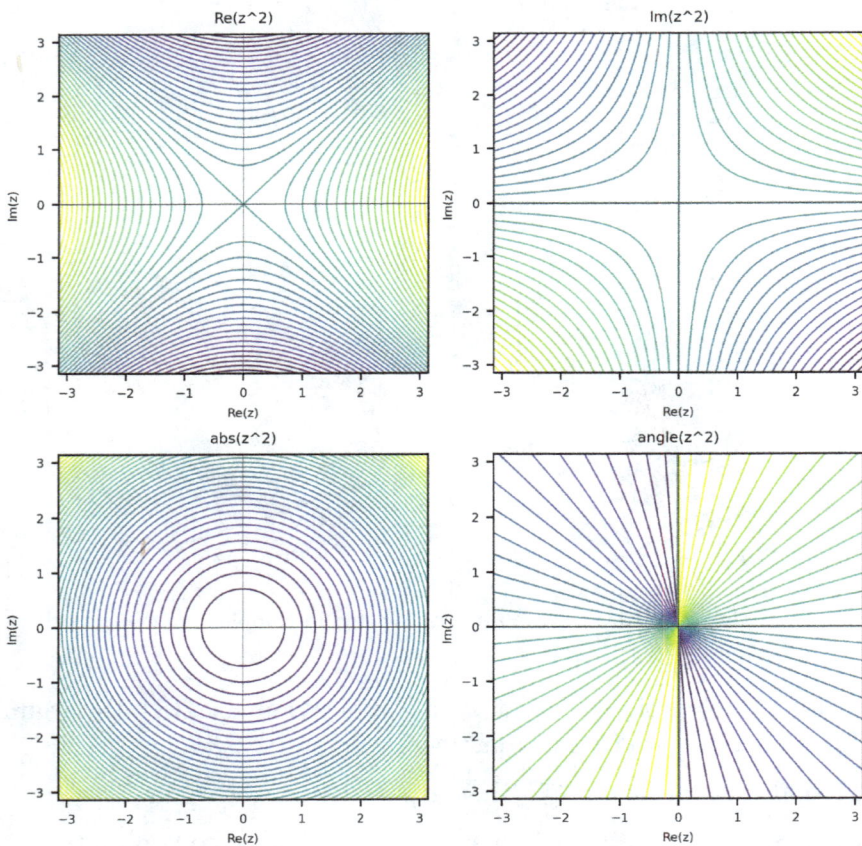

Figure 4.11.   Contour of the square function plotted in complex domain.

4.11.3 *Functions in general*

As discussed in Chapter 3, the set of $\mathbb{C}$ is closed under arithmetic, liner and nonlinear operations. Therefore, we have the following important remark:

> Elementary functions devised with arithmetic, liner and nonlinear operations are closed in $\mathbb{C}$, if the domain and codomain are set as $\mathbb{C}$, except singular pints at which the division is not defined.

This remark has been found true for all the examples we studied so far, noting that all the domains and codomains are all in $\mathbb{C}$. It will also be found true for all the examples to be given in the following discussions on various types of elementary functions.

4.12 Trigonometric functions

The trigonometric functions relate originally to right-angled triangles. These functions are induced based on the geometric relations. The independent variable is the angle between two sides of the right triangle, which in turn relates to the ratios of its side lengths. These functions are thus widely used in all sciences and engineering, and in computational methods in particular. They are typical periodic functions, have a distinct character of **periodicity**, and hence are widely used for studying periodic phenomena.

4.12.1 *Definition*

The basic trigonometric functions are sin(), cos(), and tan() (also denoted as tg()), which are familiar to readers. Their reciprocals are the cosecant, secant, and cotangent functions. Here, we introduce the sine, cosine, and tangent function as examples.

Consider the right triangle 0–a–b shown in Fig. 4.12.

The basic trigonometric functions, $\sin(\alpha)$, $\cos(\alpha)$, and $\tan(\alpha)$, are defined as

$$\sin(\alpha) = \frac{y}{r} = \frac{y}{\sqrt{x^2 + y^2}}$$

$$\cos(\alpha) = \frac{x}{r} = \frac{x}{\sqrt{x^2 + y^2}} \tag{4.15}$$

$$\tan(\alpha) = \frac{y}{x}$$

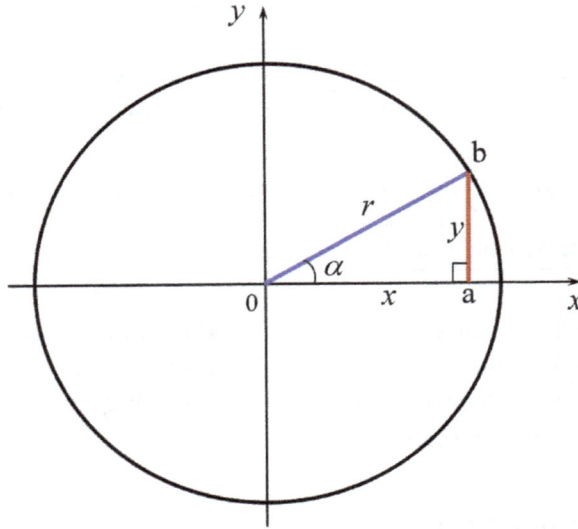

Figure 4.12. The $\sin(\alpha)$, $\cos(\alpha)$, and $\tan(\alpha)$ functions are defined using right triangle in the Cartesian coordinates.

Trigonometric functions have various applications in computational methods. It is widely used in simulations of any sinusoidal motion or phenomena, such as wave propagation and vibration [6], as well as problems involving angles and triangles. The following section describes an example problem familiar to many.

4.12.2 *Example: Sine function as the vibration modes of strings*

Music is familiar to all our readers, and many of the musical notes are produced by vibrating string. This is probably the best example to print the sine function in our mind. It presents the shape (often called mode) of a vibrating string:

$$\sin\left(\frac{n\pi x}{l}\right) \quad \text{for} \quad n = 1, 2, \ldots \tag{4.16}$$

where l is the length of the string and n stands for the nth mode. The corresponding frequency (called the nature frequency or overtones) is given by

$$f_n = \frac{nc}{2l} \tag{4.17}$$

where c is the elastic wave speed in the string. It is determined by the tension force T in the string and the line density (density per length) ρ in the form of

$$c = \sqrt{T/\rho} \tag{4.18}$$

This implies that the stiffness of a vibrating string is the tensor force in the string. We see clearly that the music is scientifically determined by three factors: (1) the tension force applied, (2) the line density and (3) the length of the string. The beauty of the music would artistically depend on how these facts are played by the musician.

The following Python code computes and plots the music wave modes that are sinusoid and harmonic:

```
1  N = 6                                      # number of wave modes
2  modes = [n for n in range(1,N+1)]
3  Labels=[str(n)+' mode' for n in modes]     # ['1st mode','2nd ',..., N]
4  xL, xR, dx = 0., 1., 0.01                        # range for the plots
5  X = np.arange(xL, xR+dx, dx)
6  l = 1.0                                     # length of the string
7
8  plt.rcParams.update({'font.size': 6})
9  fig_s = plt.figure(figsize=(6,4))
10
11 for i, n in enumerate(modes):
12     ax = fig_s.add_subplot(3,2,i+1)
13     ax.plot(X,np.sin(n*np.pi*X/l), 'b', lw=1.2,label=Labels[i])
14     ax.plot(X,-np.sin(n*np.pi*X/l),'b:')
15     if i==4 or i==5: ax.set_xlabel('x')
16     if i==0 or i==2 or i==4: ax.set_ylabel('Amplidute')
17     ax.legend()
18
19 plt.savefig('images/VibModeSting.png',dpi=500,bbox_inches='tight')
20 plt.show()
```

Figure 4.13 shows the shape of six vibration modes of a string under tension.

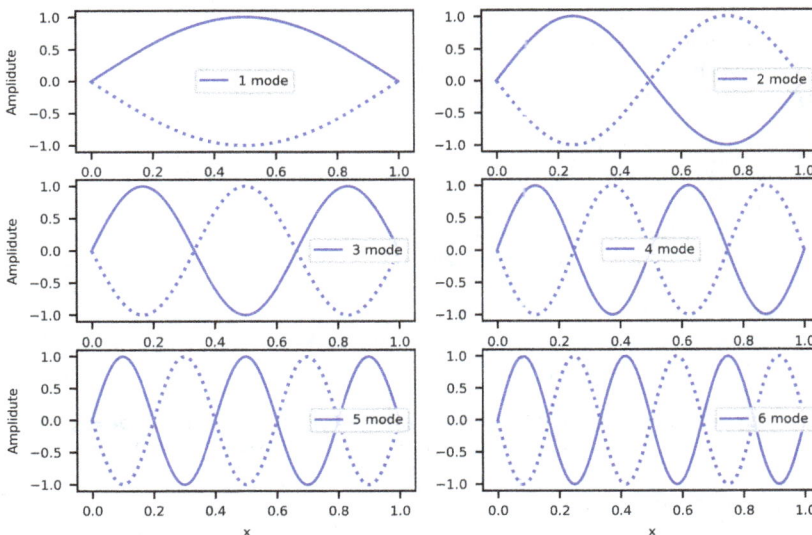

Figure 4.13. Vibration mode shape of a string under tension.

A piece of music is essentially a combination of all the wave modes.

Note that Eq. (4.16) also gives the modes of elastic waves in a bar with two ends fixed [6] or the vibration modes of a simply–simply supported beam. The only change is the corresponding wave speed c. As we have seen for multiple times, the same mathematical formula describes different types of physical phenomena.

In addition, when the boundary conditions change, the wave modes may be in the cosine function form, and these waves are still harmonic. Hence, the trigonometric functions can give good representations of them.

4.13 Closure of trigonometric functions in $\mathbb{C}$

When the domain for a trigonometric function is complex domain $\mathbb{C}$, its codomain is also $\mathbb{C}$. Trigonometric functions are closed in $\mathbb{C}$ except at the singular points.

The following code plot the distribution of sine function in a complex domain:

```python
 1  xR = np.linspace(-np.pi, np.pi, 200)   # range in real axis
 2  xI = np.linspace(-np.pi, np.pi, 200)   # range in imaginary axis
 3  lxR, lxI = len(xR), len(xI)
 4
 5  XR, XI = np.meshgrid(xR, xI)
 6  sinz = np.sin(XR + XI*1j)
 7  Z = (      sinz.real,    sinz.imag,    np.abs(sinz), np.angle(sinz))
 8  z_labels=['Re(sin(z))','Im(sin(z))','abs(sin(z))','angle(sin(z))']
 9
10  plt.rcParams.update({'font.size': 6})
11  plt.rcParams["figure.autolayout"] = True
12  fig_s = plt.figure(figsize=(6,4))
13
14  for i in range(len(Z)):
15      ax = fig_s.add_subplot(2,2,i+1, projection='3d')
16      ax.set_xlabel('Re(z)', labelpad=-11)
17      ax.set_ylabel('Im(z)', labelpad=-11)
18      ax.tick_params(axis='x', pad=-5)
19      ax.tick_params(axis='y', pad=-5)
20      ax.tick_params(axis='z', pad=-5)
21      #ax.set_zlabel(z_labels[i], rotation=0)
22      plt.title(z_labels[i], pad=-25)
23
24      ax.plot_surface(XR,XI,Z[i],color='b',rstride=1,cstride=1,
25                      shade=False,cmap="jet", linewidth=1)
26
27  fig_s.tight_layout()
28  plt.savefig('images/sinz_complex.png',dpi=500,bbox_inches='tight')
29  plt.show()
```

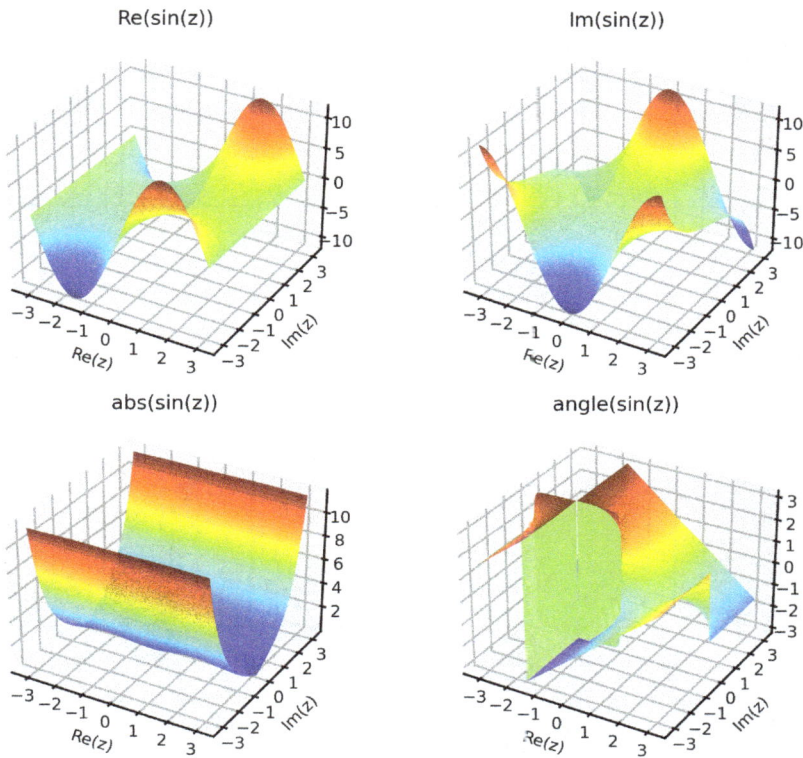

Figure 4.14. Sine function plotted in complex domain.

Figure 4.14 plots the surface of $\sin(z)$ in complex domain. Both the argument and function value are complex numbers.

It is found that for any $z \in \mathbb{C}$, the output of $\sin(z)$ is always in $\mathbb{C}$.

An alternative view is contour plots. The following code does this:

```
1  fig_s = plt.figure(figsize=(6,6))
2
3  for i in range(len(Z)):
4      ax = fig_s.add_subplot(2,2,i+1)
5      ax.set_xlabel('Re(z)'); ax.set_ylabel( Im(z)')
6      ax.axvline(x=0,c="k",lw=.3); ax.axhline(y=0,c="k",lw =.3)
7      plt.title(z_labels[i])
8
9      ax.contour(XR,XI,Z[i], levels=40, linewidths=0.8)
10
11 fig_s.tight_layout()
12 plt.savefig('images/sinz_complexC.png',dpi=500,bbox_inches='tight')
13 plt.show()
```

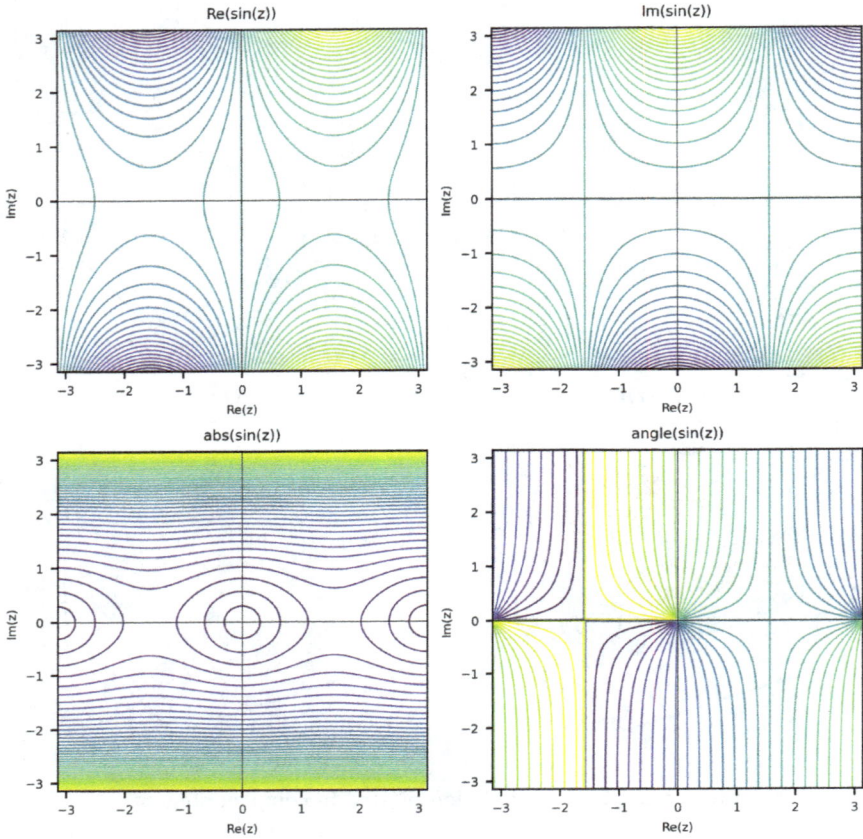

Figure 4.15. Contour of the sine function plotted in complex domain.

Figure 4.15 shows the contour plots of $\sin(z)$ in complex domain.

4.13.1 *Trigonometric identities*

Due to the geometric relations, many identities exist among trigonometric functions. The following two are examples of angle addition equations:

$$\sin(\alpha + \beta) = \sin(\alpha)\cos(\beta) + \cos(\alpha)\sin(\beta)$$

$$\cos(\alpha + \beta) = \cos(\alpha)\cos(\beta) - \sin(\alpha)\sin(\beta) \tag{4.19}$$

where α and β are used as the arguments. The following two are examples of angle subtraction:

$$\sin(\alpha - \beta) = \sin(\alpha)\cos(\beta) - \cos(\alpha)\sin(\beta)$$

$$\cos(\alpha - \beta) = \cos(\alpha)\cos(\beta) + \sin(\alpha)\sin(\beta) \tag{4.20}$$

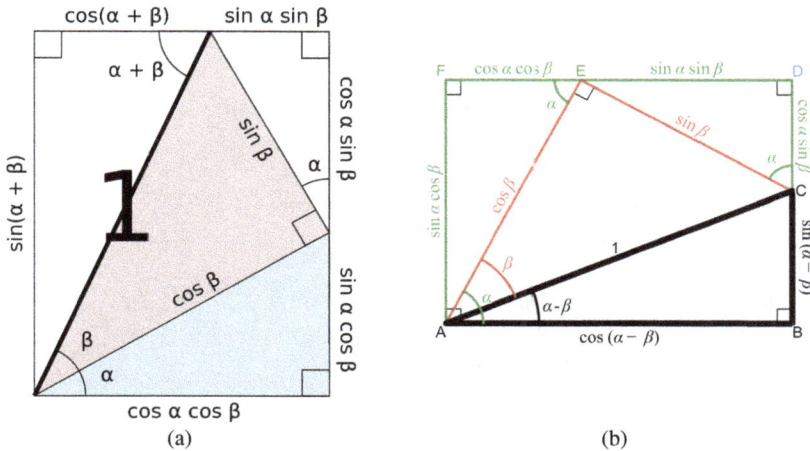

Figure 4.16. Pictorial proof of angle addition (a) and subtraction (b) formulas for the sine and cosine functions with acute angles. The black thick inclined lines have a unit length. Images are from the commons.wikimedia: (a) by Gaiacarra under the CC BY-SA 3.0 license; (b) by RajRaizada under the CC BY-SA 4.0.

A nice pictorial proof of these identities is shown in Fig. 4.16. For the angle addition equations, for example, the sine angle addition is given by the length of the left edge of the rectangle, which is the sum of two edge segments on the right, as shown in Fig. 4.16(a). The cosine angle addition is given by the length of the left segment of the top edge, which is the bottom edge minus the right segment of the top edge.

For the angle subtraction equations, Fig. 4.16(b) shows the pictorial proof.

Due to the geometric nature of the trigonometric function, the proof of many identities can often be naturally done via geometric means, as shown in Fig. 4.16. A quite complete list of trigonometric identities and proofs can be found at the wiki page.

Here, we introduce some examples of Sympy codes, so that readers can easily obtain the relationships among these functions with ease:

```
1  # example of formulas for angle sum:
2  from sympy.simplify.fu import TR6, TR9
3  α, β, π = sp.symbols("α, β, π",real=True)
4
5  print(f'sin(α+β)=',TR9(sp.sin(α+β).expand(trig=True)))
6  print(f'cos(α+β)=',TR9(sp.cos(α+β).expand(trig=True)))
```

$\sin(\alpha + \beta) = \sin(\alpha)^{*}\cos(\beta) + \sin(\beta)^{*}\cos(\alpha)$

$\cos(\alpha + \beta) = -\sin(\alpha)^{*}\sin(\beta) + \cos(\alpha)^{*}\cos(\beta)$

```
1  # example of some well-known formulas:
2  print(f'sin(α)^2+cos(α)^2=',TR6((sp.sin(α)**2+sp.cos(α)**2)))
3  print(f'cos(α)^2=',TR6(sp.cos(α)**2))
```

$\sin(\alpha)^{\wedge}2 + \cos(\alpha)^{\wedge}2 = 1$

$\cos(\alpha)^{\wedge}2 = 1 - \sin(\alpha)^{**}2$

```
1  # example of angle formulas:
2  print(f'sin(α+π/2)=',sp.sin(α+sp.pi/2))
3  print(f'cos(α+π/2)=',sp.cos(α+sp.pi/2))
```

$\sin(\alpha+\pi/2) = \cos(\alpha)$

$\cos(\alpha+\pi/2) = -\sin(\alpha)$

Readers may find detailed instructions on how to use these types of formulas in Sympy documentation.

Let us now use the following code to plot the curves for the three most fundamental trigonometric functions, $\cos(x)$, $\sin(x)$, and $\tan(x)$ in the domain of $(-2\pi, 2\pi)$:

```
1  Pi = np.pi; xL = -2*Pi; xR = 2*Pi
2  n_roots = int(2*(xR-xL)/Pi)+1
3  x = np.linspace(xL, xR, 500)              # values for array x
4  x_roots = np.linspace(xL, xR, n_roots)
5
6  fig, ax = plt.subplots(1,1,figsize=(5,3))     # Create a figure
7
8  ax.plot(x, np.cos(x), c='b', label='cos(x)')     # Plot cos(x)
9  ax.plot(x, np.sin(x), c='r', label='sin(x)')     # Plot sin(x)
10 ax.plot(x, np.tan(x), c='g', label='tan(x)')     # Plot tan(x)
11 ax.scatter(x_roots, [0]*n_roots, c='r', s=10, label='Roots')
12
13 ax.set_xlabel('x'); ax.set_ylabel('f(x)')
14 ax.set_title('Three typical trigonometric functions')
15 ax.grid(color='r', linestyle=':', linewidth=0.5)
16
17 ax.set_ylim(-5, 5)
18 ax.axvline(x=0, c="k", lw=0.6); ax.axhline(y=0, c="k", lw=0.6)
19 ax.legend()
20 plt.savefig('images/Trigonometries.png', dpi=500)
21 plt.show()
```

When the independent variable or argument x is confined in $\mathbb{R}$, the trigonometric functions are real scalar functions. The domain of trigonometric functions is $\mathbb{R}$, and codomain is in a domain in $\mathbb{R}$, depending on the

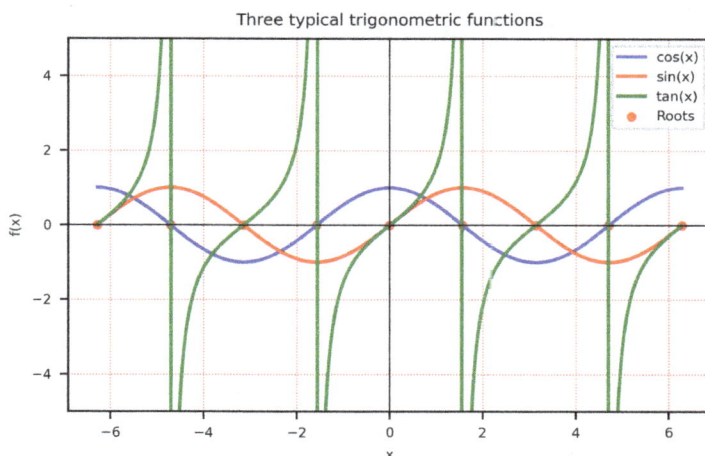

Figure 4.17. Sample trigonometric functions: sine (sinusoidal, bounded), cosine (sinusoidal, bounded), and tangent (sinusoidal, unbounded discrete points $(\pi/2 + k\pi)$, monotonic in its primary domains).

type of function. For example, the codomain for sin() is $[-1, 1] \in \mathbb{R}$ when its domain is set as $\mathbb{R}$. The codomain of tan() is $\mathbb{R}$, but the domain of tan() is $\mathbb{R}$, excluding discrete points $(\pi/2 + k\pi)$ for all integers k.

Figure 4.17 plots three example trigonometric functions: sine, cosine, and tangent.

The continuous domain for the sine and cosine functions should be $\mathbb{R}$, which can be confirmed using the following code:

```
1  from sympy.calculus.util import continuous_domain
2  continuous_domain(sp.sin(sx), sx, S.Reals)
```

$\mathbb{R}$

The continuous domain for the tangent functions is $\mathbb{R}$, but excluding these singular points:

```
1  continuous_domain(sp.tan(sx), sx, S.Reals)
```

$$\mathbb{R} \setminus \left(\left\{ 2n\pi + \frac{\pi}{2} \;\middle|\; n \in \mathbb{Z} \right\} \cup \left\{ 2n\pi + \frac{3\pi}{2} \;\middle|\; n \in \mathbb{Z} \right\} \right)$$

The major features of the three basic trigonometric functions are as follows:

1. It is a periodic function with period of 2π (for sine and cosine) and π (for tangent):

$$\sin(x + 2k\pi) = \sin x$$

$$\cos(x + 2k\pi) = \cos x \qquad (4.21)$$

$$\tan(x + k\pi) = \tan x$$

2. It has infinite number of roots that are distributed with its periodicity.
3. It is either an even function (cosine) or an odd function (for sine and tangent).

These basic properties may carry forward to other trigonometric functions and their variants.

4.13.2 *Reciprocal of trigonometric functions*

Trigonometric functions have corresponding reciprocals. For example, the secant function $\sec(x)$ is the reciprocal of $\sin(x)$: $\sec(x) = \frac{1}{\cos(x)}$. As division is involved, these types of functions can often have singularities. The singularity of $\sec(x)$ is at $x = (n + 1/2)\pi$:

```
1  import mpmath as mp
2
3  for x in [0, 0.5]:
4      print(f'mp.sec({x})    = {mp.sec(x)}')       # use sec in mpmath module
5      print(f'1/mp.cos({x}) = {1./mp.cos(x)}\n')
6
7  x = np.pi/2
8  print(f'mp.sec({x}) = {mp.sec(x)}')                # very large number
```

```
mp.sec(0)    = 1.0
1/mp.cos(0) = 1.0


mp.sec(0.5)    = 1.13949392732455
1/mp.cos(0.5) = 1.13949392732455


mp.sec(1.5707963267948966) = 1.63312393531954e+16
```

```
1  # cosecant function:
2
3  for x in [0.1, 0.5]:
4      print(f'mp.csc(x)    = {mp.csc(x)}')      # use csc in mpmath module
5      print(f'1/mp.sin({x}) = {1./mp.sin(x)}\n')
6
7  #print(f'mp.csc({0})={mp.csc(0)}')      # would give a ZeroDivisionError
```

```
mp.csc(x)    = 10.0166861316348
1/mp.sin(0.1) = 10.0166861316348


mp.csc(x)    = 2.08582964293349
1/mp.sin(0.5) = 2.08582964293349
```

```
1  #cotangent function:
2
3  for x in [0.1, 0.5]:
4      print(f'mp.cot(x)    = {mp.cot(x)}')                    # cotangent
5      print(f'1/mp.tan({x}) = {1./mp.tan(x)}\n')
6
7  #print(f'mp.cot(0) = {mp.cot(0)}')      # would give a ZeroDivisionError
```

```
mp.cot(x)    = 9.96664442325924
1/mp.tan(0.1) = 9.96664442325924


mp.cot(x)    = 1.83048772171245
1/mp.tan(0.5) = 1.83048772171245
```

4.13.3 *Inverse of trigonometric functions*

Trigonometric functions can have corresponding inverse functions. The inverse functions of sin(), cos(), and tan() are

$$\sin^{-1}(\alpha) = \arcsin(\alpha) \quad \in [-\pi/2, \pi/2]$$
$$\cos^{-1}(\alpha) = \arccos(\alpha) \quad \in [0, \pi] \qquad (4.22)$$
$$\tan^{-1}(\alpha) = \arctan(\alpha) \quad \in [-\pi/2, \pi/2]$$

Since trigonometric functions are periodic, their inverse function will naturally be multi-valued, implying that it will have multiple values for a single input. Therefore, we often use their principle values, which are shown in

Eq. (4.22). The following code plots the curves of the principle values of three basic inverse trigonometric functions:

```python
y = np.linspace(-0.999, 0.999, 500)                    # values for array x

fig, ax = plt.subplots(1,1,figsize=(3,3))              # Create a figure

ax.plot(y, np.arccos(y), c='b', label='arccos(x)')
ax.plot(y, np.arcsin(y), c='r', label='arcsin(x)')
ax.plot(y, np.arctan(y), c='g', label='arctan(x)')

ax.set_xlabel('y'); ax.set_ylabel('$f^{-1}$(y)')
ax.set_title('Inverse functions of 3 trigonometric functions')
ax.scatter(0, 0, c='r', s=10, label='Roots')

ax.grid(color='r', linestyle=':', linewidth=0.3)
ax.axvline(x=0, c="k", lw=0.6); ax.axhline(y=0, c="k", lw=0.6)
ax.legend()
plt.savefig('images/iTrigonometries.png', dpi=500)
plt.show()
```

Figure 4.18 plots the inverse functions of three trigonometric functions: sine, cosine, and tangent.

We now examine the inverse functions in extended domains including complex domain using the mpmath module.

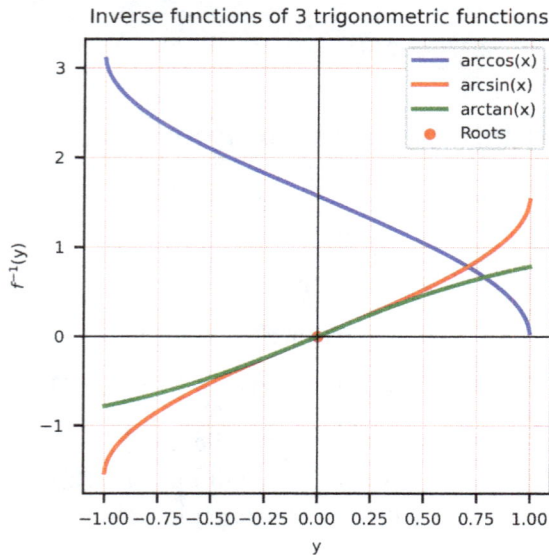

Figure 4.18.   Example of inverse trigonometric functions.

```
1  import mpmath as mp          # use mpmath moaule
2  y_r = 0.5; y_c = 0.5j
3  print(f'arcsin({y_r}) = {mp.asin(y_r)}')
4  print(f'arcsin({y_c}) = {mp.asin(y_c)}')
5  print(f'arccos({y_c}) = {mp.acos(y_c)}')
```

```
arcsin(0.5) = 0.523598775598299
arcsin(0.5j) = (0.0 + 0.481211825059603j)
arccos(0.5j) = (1.5707963267949 - 0.481211825059603j)
```

It is confirmed by the above code that the real parts of the output are within the principle values shown in Eq. (4.22).

When y values are beyond the principle values, $\mathbb{R}$ will not be able to host the inverse function. Thus, the computation switches to $\mathbb{C}$ when mpmath module is used in the computation. This can be done because $\mathbb{C}$ is closed.

The following gives some examples:

```
1  print(f'arcsin({y_r+2*np.pi}) = {mp.asin(y_r+2*np.pi)}')
2  print(f'arcsin({y_c+2*np.pi}) = {mp.asin(y_c+2*np.pi)}')
3  print(f'arccos({y_c+2*np.pi}) = {mp.acos(y_c+2*np.pi)}')
```

```
arcsin(6.783185307179586) = (1.5707963267949 - 2.6021157440236j)
arcsin((6.283185307179586+0.5j)) = (1.49037203996785 + 2.5279098471178j)
arccos((6.283185307179586+0.5j)) = (0.0804242868270425 - 2.5279098471178j)
```

When the computation is in the complex domain $\mathbb{C}$, the inverse of an inverse function should return back to original y value:

```
1  print(f'y = {mp.sin(mp.asin(y_r))}')
2  print(f'y = {mp.sin(mp.asin(y_c))}')
3  print(f'y = {mp.cos(mp.acos(y_c))}')
```

```
y = 0.5
y = (0.0 + 0.5j)
y = (6.84598372830253e-17 + 0.5j)
```

This is true regardless of the value of y when mpmath module is used in the computation:

```
1  print(f'y = {mp.sin(mp.asin(y_r+2*np.pi))} )
2  print(f'y = {mp.sin(mp.asin(y_c+2*np.pi))} )
3  print(f'y = {mp.cos(mp.acos(y_c+2*np.pi))}' )
```

```
y = (6.78318530717958 - 4.10811975847432e-16j)
y = (6.28318530717959 + 0.5j)
y = (6.28318530717959 + 0.5j)
```

```
1  print(f'y = {mp.sin(mp.asin(y_r+2*np.pi*1j))}')
2  print(f'y = {mp.sin(mp.asin(y_c+2*np.pi*1j))}')
3  print(f'y = {mp.cos(mp.acos(y_c+2*np.pi*1j))}')
```

y = (0.5 + 6.28318530717959j)

y = (0.0 + 6.78318530717959j)

y = (4.19839586524402e-16 + 6.78318530717959j)

It is seen that the trigonometric functions are closed in the complex set $\mathbb{C}$ except at singular points.

4.13.4 *The sinc(x) function*

Sometimes, we define a special function to avoid operations that may lead to unnecessary singular number and hence break down. In addition, we often need to control the boundness of function at large arguments. Such special functions can be combined with monomial functions. For example, $\text{sinc}(x)$ function is defined as

$$\text{sinc}(x) = \begin{cases} \frac{\sin(\pi x)}{\pi x}, & \text{if } x \neq 0 \\ 1, & \text{if } x = 0 \end{cases} \tag{4.23}$$

This definition follows Numpy's convention, and π is used for normalization. One may use $\text{sinc}(x/\pi)$ to obtain the unnormalized sinc function.

Note that $\text{sinc}(x)$ is an even function because both $\sin(x)$ and x are odd functions. Thus, it makes more sense to compare $\text{sinc}(x)$ with $\cos(x)$ that is also an even function. We use the following code to plot these two curves:

```
 1  Pi = np.pi; xL = -7*Pi; xR = 7*Pi
 2  n_roots = int((xR-xL)/Pi)+1
 3  x = np.linspace(xL, xR, 800)                    # values for array x
 4  x_roots = np.linspace(xL, xR, n_roots)
 5  #x = np.linspace(-7*np.pi, 7*np.pi, 800)
 6
 7  fig, ax = plt.subplots(1,1,figsize=(4,2.5))         # Create a figure
 8
 9  ax.plot(x, np.cos(x), c='g', label='$\cos(x)$', lw=.6)  # plot cos(x)
10  ax.plot(x, np.sinc(x/np.pi), c='b', label='sinc$(x/\pi)$')  # sinc(x)
11  ax.scatter(x_roots, [0]*n_roots, c='r', s=10, label='Roots')
12
13  ax.set_xlabel('x'); ax.set_ylabel('f(x)')
14  ax.set_title('sinc(x) vs. cos(x)')
15  ax.grid(color='r', linestyle=':', linewidth=0.3)
16
17  ax.axvline(x=0, c="k", lw=0.6); ax.axhline(y=0, c="k", lw=0.6)
18  ax.legend()
19  plt.tight_layout()
20  plt.savefig('images/sincNcos.png', dpi=500)
21  plt.show()
```

Figure 4.19 plots the comparison curves of $\text{sinc}(x)$ and $\cos(x)$.

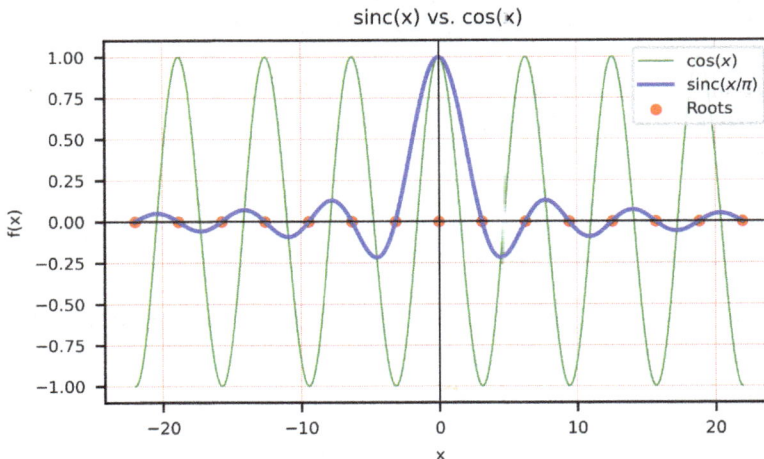

Figure 4.19. Comparison of $\text{sinc}(x)$ with $\cos(x)$.

It is seen that $\text{sinc}(x)$ has some sinusoidal behavior inherited from $\sin(x)$, but converges at $x = \pm\infty$ due to the inverse monomial function $\frac{1}{x}$, while $\cos(x)$ is sinusoidal, bounded, but does not converge at $x = \pm\infty$).

The $\text{sinc}(x)$ function can also take complex variables as its arguments. In this case, the output is in general complex-valued, as shown in the following codes:

```
1  for x in [0, 1, 100, 1+3j, -2+3j]:
2      print(f'sinc({x:.2f}) = {np.sinc(x):.2f}')
```

```
sinc(0.00) = 1.00
sinc(1.00) = 0.00
sinc(100.00) = 0.00
sinc(1.00+3.00j) = -591.66-197.22j
sinc(-2.00+3.00j) = 455.12-303.41j
```

```
1  x = 1.0
2  print(f'Normalized sinc    = {np.sinc(x):.4e}')
3  print(f'Unnormalized sinc = {np.sinc(x/np.pi):.4e}')
4  print(f'Unnormalized sinc = {(np.sin(x)/x):.4e}')
```

```
Normalized sinc    = 3.8982e-17
Unnormalized sinc = 8.4147e-01
Unnormalized sinc = 8.4147e-01
```

4.13.5 *The $\sin(1/x)$ function*

The composition of trigonometric and monomial functions can lead to functions with very interesting behavior. The following function is a very well-known one:

$$f(x) = \sin\frac{1}{x} \tag{4.24}$$

It is undefined at $x = 0$. We use the following code to plot this curve:

```
1  x = np.linspace(-1.*np.pi, 1.*np.pi, 2000)  # values for array x
2
3  fig, ax = plt.subplots(1,1,figsize=(4,2.5))            # new figure
4
5  ax.plot(x, np.sin(x), c='b', label='$\sin(x)$', lw=.3)  # cos(x)
6  ax.plot(x, np.sin(1/x), c='r', label='$\sin(1/x)$', lw=.3)
7
8  ax.set_xlabel('x'); ax.set_ylabel('f(x)')
9  ax.set_title('$\sin(1/x)$')
10 ax.grid(color='r', linestyle=':', linewidth=0.2)
11
12 ax.axvline(x=0, c="k", lw=0.6); ax.axhline(y=0, c="k", lw=0.6)
13 ax.legend()
14 plt.tight_layout()
15 plt.savefig('images/sin1_over_x.png', dpi=500)
16 plt.show()
```

Figure 4.20 plots the curves of $\sin(1/x)$ together with $\sin(x)$ for comparison. They are both odd functions.

It is seen that when x is in the vicinity of zero, the rational function $\frac{1}{x}$ changes vary fast. This is compounded with the periodicity of sine function leading to violent oscillation. When x approaches infinity, the rational function $\frac{1}{x}$ approaches zero, and hence, the entire function converges to zero.

```
1  from sympy.calculus.util import continuous_domain
2  continuous_domain(sp.sin(1/sx), sx, S.Reals)
```

$(-\infty, 0) \cup (0, \infty)$

This function is found continuous in domain $\mathbb{R}$, excluding $x = 0$.

4.13.6 *Euler's formula for trigonometric functions*

Both sine and cosine relate to the exponential function (to be discussed in the following section), which is known as Euler's formula (also given in Section 3.2.3 in a slightly different form):

$$\sin x = \frac{e^{ix} - e^{-ix}}{2i}; \cos x = \frac{e^{ix} + e^{-ix}}{2} \qquad (4.25)$$

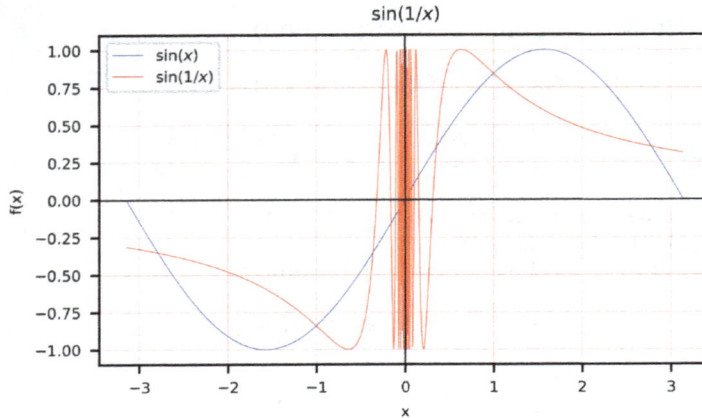

Figure 4.20. Comparison of $\sin(1/x)$ and $\sin(x)$. This function has infinite number of roots in the vicinity of $x = 0$.

This implies that trigonometric functions can have complex arguments and their values can also be complex-valued, as shown in the following example. This important formula connects the trigonometric and exponential functions: they meet in the complex domain:

```
1  x_r = 0.5; x_c = 0.5j
2  print(f'sin(x) with real argument = {np.sin(x_r):.4f}')
3  print(f'Euler= {((np.exp(x_r*1j)-np.exp(-x_r*1j))/2j):.4f}\n')
4  print(f'sin(x) with complex argument = {np.sin(x_c):.4f}')
5  print(f'Euler= {((np.exp(x_c*1j)-np.exp(-x_c*1j))/2j):.4f}\n')
6  print(f'cos(x) with complex argument = {np.cos(x_c):.4f}')
7  print(f'Euler= {((np.exp(x_c*1j)+np.exp(-x_c*1j))/2):.4f}')
```

```
sin(x) with real argument = 0.4794
Euler= 0.4794+0.0000j
```

```
sin(x) with complex argument = 0.0000+0.5211j
Euler= 0.0000+0.5211j
```

```
cos(x) with complex argument = 1.1276-0.0000j
Euler= 1.1276+0.0000j
```

4.14 Polynomial functions

4.14.1 *General form*

One of the most widely used classes of functions is the polynomial functions. It can be generally expressed as

$$f(x) := c_0 + c_1 x + c_2 x^2 + \cdots + c_r x^n = \sum_{k=0}^{n} c_k x^k \qquad (4.26)$$

where x is the independent number variable and $c_k, k = 0, 1, 2, \ldots, n$ are given numbers called coefficients. The number of coefficients is $n + 1$. They are all in general defined in $\mathbb{C}$, with n being an integer number. Polynomials are also specifically denoted as $P(x)$. The domain and codomain of $f(x)$ are $\mathbb{C}$. Since the highest power of x is n, $f(x)$ is said to be n-order polynomial. When the coefficients for all terms below including nth-order terms, are non-zero, the polynomial is said to have a complete order n. Equation (4.7) is a first-order polynomial and is a special case of Eq. (4.26). The square function given in Eq. (4.3) is also a special case of Eq. (4.26).

Equation (4.26) is formed by using **monomials**, $1, x, x^2, \ldots, x^n$, discussed in the section on monomials. These $n + 1$ monomials are linearly independent and hence form general polynomial function expression (4.26). Any polynomial of order n can be expressed in Eq. (4.26) by adjusting these coefficients.

4.14.2 *Major properties*

Polynomials have the following important features:

1. An nth polynomial can have up to $n + 1$ terms.
2. A polynomial equation of order n always has n roots if we count repeated and complex roots. This feature can be used to create polynomials except the constant one.
3. Finding the root of a linear polynomial is straightforward, and the formula is given in Eq. (4.9).
4. As discussed in the previous chapter, we have simple formulas to calculate two roots of a quadratic polynomial, which is given in Eq. (3.22).

5. There are also formulas for the cubic and quartic equations. These formulas are quite complicated and not used often for the purpose of root finding.
6. Based on the Abel–Ruffini theorem, no general formula exists for roots of higher degrees. We often use root-finding algorithms to find the roots numerically.

The computation of a polynomial can be done more efficiently, in terms of the number of arithmetic operations, if Eq. (4.26) is rewritten in the following nested form:

$$c_0 + x(c_1 + x(c_2 + \cdots + x(c_{n-1} + c_n x^n))) \tag{4.27}$$

This calculation can be down using the code in the following section.

4.14.3 *Numpy codes and examples for polynomials*

```python
# note the arrangement-order of the coefficients
def poly_f(coefs, x, reverse_c = False):

    '''Compute the value at x of a polynomial with its coefficients
        given. The coefficients are arranged in the order of
        c[0] + c[1]*x +... + c[n-2]*x**(n-2) + c[n-1]*x**(n-1)
        If arrangement is reversed, set reverse_c=True.
        The highest order:n-1

    Inputs:
        coefs: list;
        x    : independent variable, float, integer or array.
    Return:
        poly_value: same shape as x.
    '''
    if reverse_c: coefs.reverse(); print(f'coefs:{coefs}')
    poly_value = 0; xk =1

    for c in coefs:
        poly_value += c*xk    # add upon the next term
        xk = x*xk             # xk is x^k
    return poly_value
```

It is seen clearly that once the coefficients of the polynomial is given, its value is determined. Using the coefficients, one can create all possible polynomials, including the constant one.

The following code cell computes the values of two polynomials for given coefficients using our poly_f() method:

```
1  np.random.seed(8)
2  x = np.random.randn(2)          # random x for testing
3
4  coefs3 = [1, 2, 3]
5  coefs4 = [5, 4, 3, 2, 1]
6
7  print(f'at x = {x}, poly_value={poly_f(coefs3, x, reverse_c=True)}')
8  print(f'at x = {x}, poly_value={poly_f(coefs4, x, reverse_c=True)}')
```

```
coefs:[3, 2, 1]
at x = [0.09120472 1.09128273], poly_value=[3.19072773 6.37346347]
coefs:[1, 2, 3, 4, 5]
at x = [0.09120472 1.09128273], poly_value=[ 1.21074498 19.04487547]
```

Numpy has a polyval() method to compute polynomial values at given x. It also needs just a list of coefficients. Attention is needed on the arrangement order of these coefficients required:

```
1  # Note the order used in np.polyval:
2  # c[0]*x**(n-1) + c[1]*x**(n-2)+...+c[n-2]*x + c[n-1]
3
4  coefs3.reverse()                # reverse the order of coefs
5  coefs4.reverse()
6
7  print(f'at x = {x}, poly_value={np.polyval(coefs3, x)}')
8  print(f'at x = {x}, poly_value={np.polyval(coefs4, x)}')
```

```
at x = [0.09120472 1.09128273], poly_value=[3.19072773 6.37346347]
at x = [0.09120472 1.09128273], poly_value=[ 1.21074498 19.04487547]
```

In Numpy, one can also create polynomials by specifying the roots of it:

```
1  # Create a polynomial by giving a set of roots
2  roots = [-3, -2, -.5, 2, 4]
3  coefs = np.poly(roots)          # compute coefficients using roots
4                                  # use help(np.poly) to find more
5  print(coefs)
6  f_poly = np.poly1d(coefs)       # produce the polynomial function
7                                  # use help(np.poly1d) to find more
8  print(f_poly)
```

```
[  1.    -0.5 -16.5  -4.    50.    24. ]
   5      4       3      2
1 x - 0.5 x - 16.5 x - 4 x + 50 x + 24
```

As a special case, if no root is given, np.poly() gives the coefficient for constants, which is 1.0:

```python
root0 = []                              # Create a constant
coef1 = np.poly(root0)           # compute coefficients using roots
                                       # use help(np.poly) to find more
print(coef1)
f_poly1 = np.poly1d(coef1)         # produce the polynomial function
                                       # use help(np.poly1d) to find more
print(f_poly1)
```

```
1.0
```

```
1
```

```python
x = sp.symbols('x')
f_poly(x)
```

$$x\left(x\left(x\left(x\left(1.0x - 0.5\right) - 16.5\right) - 4.0\right) + 50.0\right) + 24.0$$

To have the polynomial printed nicely and with its domain specified, one may use Sympy:

```python
sp_poly = sp.Poly.from_list(coefs, sp.symbols('x'))
sp_poly
```

$$\text{Poly}\left(1.0x^5 - 0.5x^4 - 16.5x^3 - 4.0x^2 + 50.0x + 24.0, x, domain = \mathbb{R}\right)$$

For a given high-order polynomial, we do not usually know its roots. The np.roots() method can be used to find them if its coefficients are given:

```python
roots = np.roots(coefs)              # find the roots of the polynomial
print(f"The roots of the polynomial are: {roots}")

f_poly.r                                  # or simply use .r
```

```
The roots of the polynomial are: [ 4.   2.  -3.  -2.  -0.5]
```

```
array([ 4. ,  2. , -3. , -2. , -0.5])
```

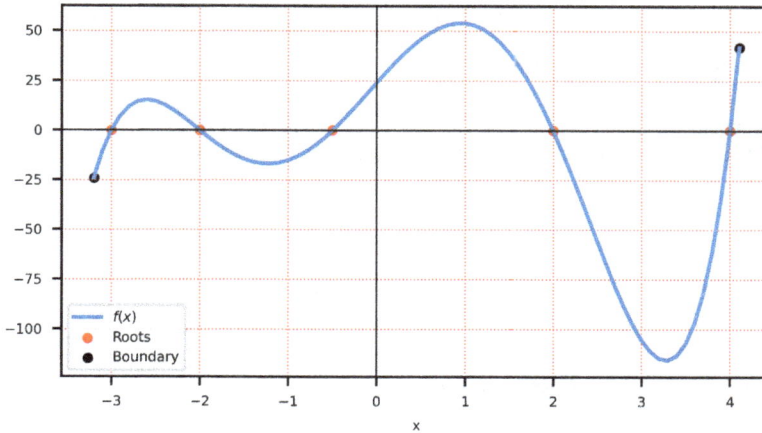

Figure 4.21. A fifth-order polynomial.

These roots are the same as those given earlier. The only difference may be the arrangement of these roots.

For a given Numpy polynomial object, finding its coefficients is easy:

```
1  f_poly.c                      # simply use .c
```

```
array([  1. ,   -0.5, -16.5,   -4. ,   50. ,   24. ])
```

In the following Numpy code, we print the roots and the boundary points of the domain together with the curve of the polynomial given above:

```
 1  plt.rcParams.update({'font.size': 5})
 2  fig, ax = plt.subplots(1,1,figsize=(4,2.5))
 3  X = np.arange(-3.2, 4.2, .1)
 4  fX = poly_f(list(coefs), X, reverse_c=True)     # or use np.polyval()
 5
 6  ax.plot(X, fX, label="$f(x)$")
 7  ax.scatter(roots,np.zeros(len(roots)), c='r', s=8, label="Roots")
 8  ax.scatter([X[0],X[-1]],[fX[0],fX[-1]],c='k', s=8, label="Boundary")
 9
10  ax.set_xlabel('x')
11  ax.set_title(f'A {len(roots)}th order polynomial')
12  ax.grid(color='r', linestyle=':', linewidth=0.5)
13  ax.axvline(x=0, c="k", lw=0.6); ax.axhline(y=0, c="k", lw=0.6)
14  ax.legend(loc='lower left')
15
16  plt.savefig('images/p5th.png', dpi=500)
17  plt.show()
```

```
coefs:[24.0, 50.0, -4.0, -16.5, -0.5, 1.0]
```

Figure 4.21 plots the curve of a fifth-order polynomial with its five roots. The following code checks the continuous domain of this polynomial:

```
1  from sympy.calculus.util import continuous_domain
2  continuous_domain(sp.Poly(coefs, sx), sx, S.Reals)
```

$\mathbb{R}$

This function is found continuous in domain $\mathbb{R}$.

4.14.4 *Quadratic functions*

Among the polynomial functions, two functions are very special and most often used in computational methods. One is the linear function discussed earlier and another is the quadratic function that is a second-order polynomial. The latter has the following general form:

$$f(x) = c_0 + c_1 x + c_2 x^2 \tag{4.28}$$

where c_0 and c_1 are arbitrary numbers and c_2 is a non-zero number. They can be in $\mathbb{C}$. When c_2 is zero, the square function becomes linear. When $c_0 = c_1 = 0$ and $c_2 = 1$, it is the square function we discussed in the beginning of this chapter. A quadratic function has two roots that can be computed using the quadratic formula.

The following code computes and plots the roots of a given quadratic function. We also compute its extreme value, which is located at the root of the derivative of the quadratic function. We will discuss about the derivative of a function in greater detail in a separated volume. Here, we simply use it:

```
1  # define an arbitrary quadratic function in numpy
2  def f_quad(x, c0, c1, c2):
3
4      '''Define an arbitrary quadratic function given coefficients.
5      All its key values, extreme & 2 roots are computed.
6      '''
7      f  = c2*x**2 + c1*x + c0
8      xm = -c1/c2/2.                      # location of the extreme value
9      fm = c2*xm**2 + c1*xm + c0                    # the extreme value
10
11     bac=np.sqrt(c1**2-4*c2*c0)
12     root1 = .5*(-c1+bac)/c2          # roots, use the quadratic formula
13     root2 = .5*(-c1-bac)/c2
14
15     return f, xm, fm, (root1, root2)
```

Plot a typical quadratic function using the following code:

```
 1  plt.rcParams.update({'font.size': 5})
 2  fig, ax = plt.subplots(1,1,figsize=(3,2))
 3  xL, xR = -7.5, 5
 4  X = np.arange(xL, xR, .01)
 5
 6  c0, c1, c2 = 8., 5, -2
 7  yL, *_ = f_quad(xL, c0, c1, c2)
 8  yR, *_ = f_quad(xR, c0, c1, c2)
 9
10  fq, xm, fm, roots = f_quad(X, c0, c1, c2)
11
12  ax.plot(X, fq, label="$f(x)$")
13  ax.scatter(xm, fm, c='m', s=8, label="Extreme")
14  ax.scatter(roots, [0]*len(roots), c='r', s=8, label="Roots")
15  ax.scatter([xL, xR], [yL, yR], c='k', s=8, label="Boundaries")
16
17  ax.set_xlabel('x')
18  ax.set_title('Quadratic function')
19  ax.grid(color='r', linestyle=':', linewidth=0.3)
20  ax.axvline(x=0, c="k", lw=0.6); ax.axhline(y=0, c="k", lw=0.6)
21  ax.legend(loc='lower right')
22
23  plt.savefig('images/f-quad.png', dpi=500)
24  plt.show()
```

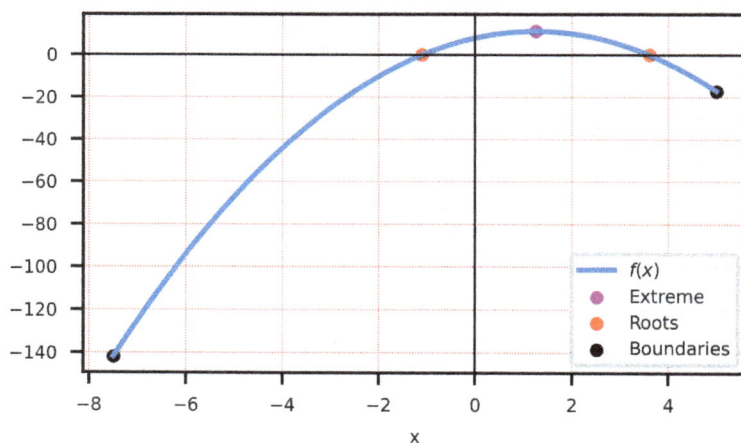

Figure 4.22. A quadratic function. It can only have one extreme (maximum in this case) value.

4.14.5 *Inverse of a polynomial function*

The inverse of a polynomial function can be expressed as follows:

$$x = f^{-1}(y) \tag{4.29}$$

For linear polynomials, the inverse function is given by Eq. (4.11). For quadratic polynomials, it can be found by finding the roots of the following equation:

$$y = c_0 + c_1 x + c_2 x^2 \tag{4.30}$$

The two roots of Eq. (4.30) can be found using the known as the quadratic formula:

$$x = \frac{-c_1 \pm \sqrt{c_1^2 - 4c_2(c_0 - y)}}{2c_2} \tag{4.31}$$

which gives two equations that can always be found so long as $c_2 \neq 0$. We can assume that c_2 is non-zero. Otherwise, the polynomial is a linear function, and we have the inverse function given in Eq. (4.11).

Equation (4.31) gives two solutions that are in general complex. This implies that there are multiple inverse functions for a given polynomial function. In general, there should be n inverse functions for an n-order polynomial, counting repeated and complex ones. These inverse functions can be found numerically, especially for polynomials with order larger than 2. The procedure should be as follows.

First, find the roots of the modified polynomial expression for a given value of y:

$$\underbrace{(c_0 - y)}_{c_{0y}} + c_1 x + c_2 x^2 + \cdots + c_n x^n = 0 \tag{4.32}$$

These roots in x become dependent variables for a given value of the inverse function y. Therefore, when y varies over its domain, we will have multiple pieces of functions, each of which corresponding to a root. The number for the inverse function pieces should be the same as the number of roots of the original function. This is quite typical when dealing with inverse problems [3].

4.14.6 *Python code to generate the inverse function of a polynomial*

Let us write the following codes to generate and plot inverse functions of polynomials. We consider two cases as examples: the simple square function given in Eq. (4.3) and the fifth-order polynomial shown in Fig. 4.21.

The following code function computes x values for given y by finding the roots of Eq. (4.32) for any polynomial with given coefficients:

```python
def y_to_x(coefs, c0, y):

    '''Find the x values (that is the real part of the roots of the
        modified polynomial) for given an y value.
    '''
    coefsy = coefs.copy()
    coefsy[-1] = c0-y                          # replace the constant
    xs = np.roots(coefsy)             # find the roots of the polynomial
    return np.sort(np.real(xs))              # take only the real part
```

Now, we use the function to generate the inverse function of the polynomial given earlier:

```python
# Consider the a polynomial with its roots known:

roots = [0, 0]                          # roots for the square function
#roots = [-3, -2, -.5, 2, 4]        # roots for the 5th order polynomial

coefs = np.poly(roots)        # produce the coefficients of the polynomial

c0 = coefs[-1].copy()                 # get the constant of the polynomial
y = np.arange(-125, 125, .1)         # domain for the inverse function y.

# Find the real part of the roots of the modified polynomial with y
X = y_to_x(coefs, c0, y[0])
for i in range(1, len(y)):
    Xr = y_to_x(coefs, c0, y[i])
    X = np.vstack((X, Xr))

plt.rcParams.update({'font.size': 5})
fig, ax = plt.subplots(1,1,figsize=(3.3,3))

# plot the y-X curves, one curve per root:
for i in range(len(roots)):
    ax.plot(y, X[:,i], alpha=0.6, label=f"x-{i}")

# plot the roots
ax.scatter(np.zeros(len(roots)), roots, c='r', s=8, label="Roots")

# plot boundary points
Xmin = np.min(X); Xmax = np.max(X)            # X values on the boundary
yb1 = np.polyval(coefs, Xmin)                 # y values on the boundary
yb2 = np.polyval(coefs, Xmax)
ax.scatter([yb1,yb2],[Xmin,Xmax], c='k', s=8, label="Boundary")

ax.set_ylabel('x'); ax.set_xlabel('y')
ax.set_title(f'Inverse functions of a {len(roots)}th order polynomial')
ax.grid(color='r', linestyle=':', linewidth=0.5)
ax.axvline(x=0, c="k", lw=0.6); ax.axhline(y=0, c="k", lw=0.6)
```

```
38  ax.legend(loc='upper left')                      # for the square function
39  #ax.Legend(loc='center left')             # for the 5th order polynomial
40
41  plt.savefig('images/p2nd_inverse.png', dpi=500)        # square function
42  #plt.savefig('images/p5th_inverse.png', dpi=500)  # 5th order polynomial
43
44  plt.show()
```

Figure 4.23 shows the results obtained for the square function in the inversion process. There always exist two x values (real part of the roots) for any given y (the horizontal lines are duplicated). This is because our computation is done in complex numbers, and the complex space is closed algebraically, as discussed in Chapter 3. It is found that there is one branching point at which two curves are joined together. This is because there are two roots of the polynomial changes between the real-valued and the complex-valued at the branching point.

Figure 4.24 shows the results obtained for the fifth-order polynomial. The is obtained using the same code by uncommenting the code lines for fifth-order polynomial. Since each of these four nearly horizontal lines is duplicated, there always exist five x values (real part of the roots) for any given y. This is because our computation is done in complex numbers, and the complex space is closed algebraically. There are four branching points at which three (in fact four) curves are joined together because the roots of

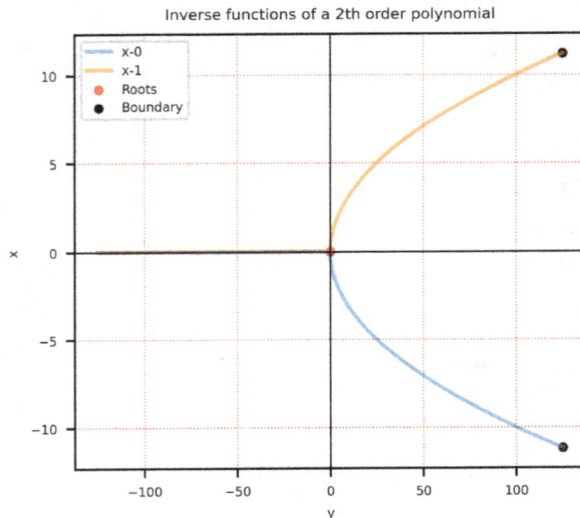

Figure 4.23. Inverse curves of the square function (simplest quadratic function). For any given y, there always exist two x values (real part of the roots), noting that the horizontal line is duplicated.

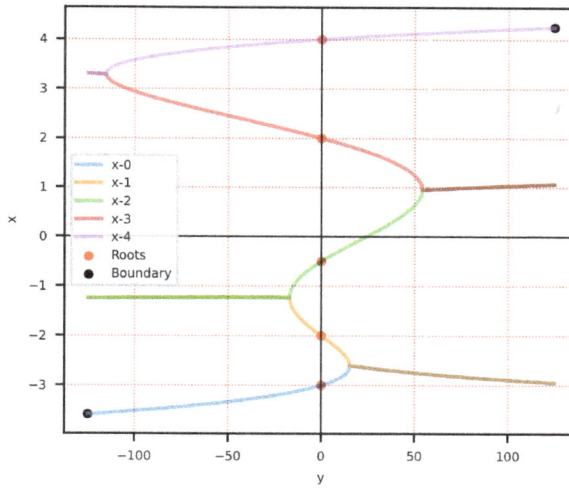

Figure 4.24. Inverse curves of the fifth-order polynomial. For any given y, there always exist five x values (real part of the roots), noting that each of these four nearly horizontal lines is duplicated at the branching points located at $\frac{dx}{dy} = \infty$.

the polynomial change between the real-valued and complex-valued at these branching points. These branching points are always located at $\frac{dx}{dy} = \infty$ (or $\frac{dy}{dx} = 0$), and critical for inverse problems. Finding inverse solutions can fail at these points [1].

If we would like to find the real inverse function, we shall remove the branches corresponding to the complex roots. We write the following code to clear this up and produce a real inverse function:

```
1   # Consider the a polynomial with its roots known:
2   roots = [-3, -2, -.5, 2, 4]                # roots for the polynomial
3   n = len(roots)                             # order of the polynomial
4   coefs = np.poly(roots)          # produce the coefficients of the poly
5   print(f"The coefs of the polynomial: {coefs}")
6
7   def invers_of_poly(coefs, b, yi):
8       coefs[-1] = b-yi
9       x_roots = np.sort(np.roots(coefs)) # find the roots and sort
10      real_roots = x_roots[np.isclose(x_roots.imag, 0)].real
11      return (real_roots, yi)            # pair with y
12
13  plt.rcParams.update({'font.size': 5})
14  fig, ax = plt.subplots(1,1,figsize=(3.3,3))
15
16  ax.set_ylabel('x'); ax.set_xlabel('y')
17  ax.set_title(f'Inverse functions of a {n}th order polynomial')
18  ax.grid(color='r', linestyle=':', linewidth=0.5)
19  ax.axvline(x=0, c="k", lw=0.6); ax.axhline(y=0, c="k", lw=0.6)
20
```

```
21  b = coefs[-1].copy()
22  y = np.arange(-125, 125, .1)
23  for i in range(len(y)):
24      X, yi = invers_of_poly(coefs, b, y[i])
25      ax.scatter([yi]*len(X), X, s=0.1, c='b')
26
27  O = np.zeros(n)
28  ax.scatter(O, roots, c='r', s=5, label="Roots of the polynomial")
29
30  ax.legend(loc='center left')#, bbox_to_anchor=(1, 0.5))
31  plt.savefig('images/p5th_inv_real.png', dpi=500)
32  plt.show()
```

The coefs of the polynomial: [1.   -0.5 -16.5 -4.   50.   24.]

Figure 4.25 gives a single (numerically) piece of inverse function of the fifth-order polynomial in real space. The inverse function is indeed multi-valued: each y value may correspond to multiple real x.

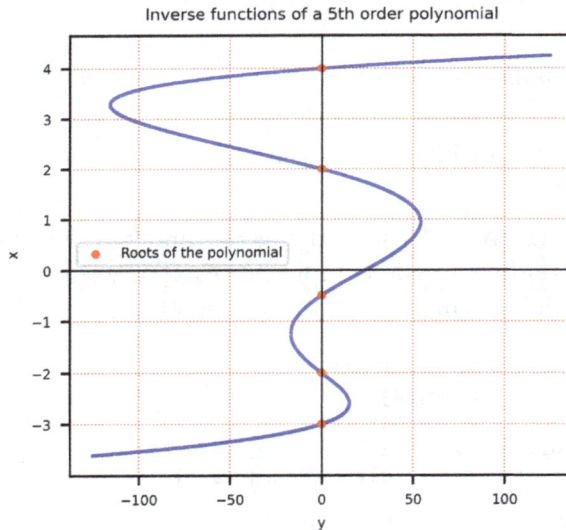

Figure 4.25.   Cleaned-up one-piece inverse curves of a fifth-order polynomial.

4.15   Rational functions

4.15.1   *Definition*

Rational function can be expressed as

$$f(x) = \frac{P(x)}{Q(x)} \quad \text{for} \quad Q(x) \neq 0 \tag{4.33}$$

where the numerator $P(x)$ and the denominator $Q(x)$ are both polynomial functions of $x \in \mathbb{C}$. They can both have real or complex coefficients and do not have any common polynomial factor except 1.

4.15.2 *Properties*

A rational function is singular at all the roots of $Q(x)$, including the complex roots. For example, it seems that the following function

$$f(x) = \frac{x^2 + 2x + 8}{x^2 + 1} \tag{4.34}$$

does not seem to have singularity, but it has one at $x = i$:

$$f(i) = \frac{i^2 + 2i + 8}{i^2 + 1} = \frac{2i + 7}{0} \tag{4.35}$$

The evaluation and behavior of rational functions can be significantly different from those of polynomial functions. This is because of the presence of the roots of $Q(x)$. If $Q(x)$ has an order of q, there will always be q roots and hence singularities (in the complex plane). These singularity points disrupt the continuity. Thus, rational functions can vary drastically in the vicinity of these singularity points.

4.15.3 *Example of a converging rational function*

Consider the following simple rational function:

$$f(x) = \frac{x^2 + x + 8}{(x^2 - 9)(x - 1)} \tag{4.36}$$

We use the following code to plot its distribution along the real axis:

```
1  coef_P = [1, 1, 8]              # define coefs
2  Px = np.poly1d(coef_P)         # produce the poly., given coefs
3  roots_P = np.roots(coef_P)
4
5  gr.printx('roots_P')
6  gr.printx('Px')
7  print()
8
9  roots_Q = [-3, 3, 1]           # define roots
10 coef_Q = np.poly(roots_Q)      # find coefs
11 Qx = np.poly1d(coef_Q)         # produce the poly., given coefs
12
13 gr.printx('roots_Q')
14 gr.printx('Qx')
```

```
roots_P = array([-0.5+2.7839j, -0.5-2.7839j])
Px = poly1d([1, 1, 8])

roots_Q = [-3, 3, 1]
Qx = poly1d([ 1., -1., -9.,  9.])
```

As shown, $P(x)$ is one order lower than $Q(x)$, and $Q(x)$ has three roots.

```
1  def ration_f(P, Q, x):
2      return P(x)/Q(x)
3
4  label_f = r"$\frac{x^2+x+8}{(x^2-9)(x-1)}$"
5  X = np.arange(-7, 7, .01)
6
7  plt.rcParams.update({'font.size': 5})
8  fig, ax = plt.subplots(1,1,figsize=(3.5,2.5))
9
10 ax.plot(X, ration_f(Px, Qx, X), label=label_f, lw=1.)
11
12 ax.set_title('Rational function with singularities')
13 ax.set_xlabel('x'); ax.set_ylim(-5, 5)
14 ax.grid(color='r', linestyle=':', linewidth=0.3)
15 ax.axvline(x=0, c="k", lw=0.4); ax.axhline(y=0, c="k", lw=0.4)
16 ax.scatter([-3,1,3],[0,0,0], c='r', s=8, label="singular")
17 ax.legend(loc='upper left')
18
19 plt.savefig('images/ration_f2.png', dpi=500); plt.show()
```

Figure 4.26 plots a rational function with three singular points. It converges when x approaches infinities.

```
1  from sympy.calculus.util import continuous_domain
2  sx = sp.symbols('sx')
3  continuous_domain(ration_f(Px, Qx, sx), sx, S.Reals)
```

$(-\infty, -3.0) \cup (-3.0, 1.0) \cup (1.0, 3.0) \cup (3.0, \infty)$

This rational function is found continuous in domain $\mathbb{R}$, excluding those singularity points.

In the above case, because the order of the numerator polynomial is lower than that of the denominator, the rational function converges to zero when x approaches ∞ or $-\infty$.

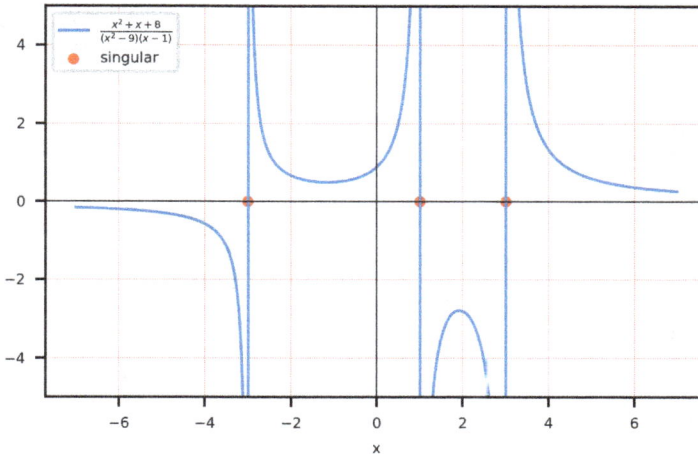

Figure 4.26. A rational function with singularities. It converges at the infinities.

```
1  from sympy import limit, oo, Symbol
2  sx = Symbol('sx')
3  print(f'Limit at left: {limit(ration_f(Px, Qx, sx), sx, -oo)}')
4  print(f'Limit at right:{limit(ration_f(Px, Qx, sx), sx, oo)}')
```

Limit at left: 0
Limit at right:0

```
1  print(f'Limit at the roots (1st, e.g.) of denominator:'
2        f' {limit(ration_f(Px, Qx, sx), sx, roots_Q[0])}')
```

Limit at the roots (1st, e.g.) of denominator: oo

4.15.4 *Example of a diverging rational function*

When the order of the numerator is higher than that of the denominator, the rational function diverges when x approaches ∞ or $-\infty$. Let take a look at the following example:

$$f(x) = \frac{x^4 - x^3 + x - 5}{(x^2 - 9)(x - 1)} \tag{4.37}$$

Its distribution along the real axis is plotted using the following code:

```python
#np.set_printoptions(precision=1, suppress=True)
coef_P = [1, -1, 0, 1, -5]                          # define coefs
Px = np.poly1d(coef_P)                # produce the poly., given coefs
roots_P = np.roots(coef_P)

print(f'All roots:')

for i in range(len(roots_P)):
    print(f'    {roots_P[i]:.4f} = {(roots_P[i]):.4f}')

roots_r_P = roots_P[np.isclose(roots_P.imag, 0)].real

np.set_printoptions(precision=4, suppress=True)
gr.printx('roots_r_P')
```

```
All roots:
    1.6883+0.0000j = 1.6883+0.0000j
    0.3500+1.4181j = 0.3500+1.4181j
    0.3500-1.4181j = 0.3500-1.4181j
   -1.3882+0.0000j = -1.3882+0.0000j
roots_r_P = array([ 1.6883, -1.3882])
```

It is found that the denominator polynomial has two real roots.

```python
label_f = r"$\frac{x^4-x^3+x-5}{(x^2-9)(x-1)}$"
X = np.arange(-9, 9, .01)

plt.rcParams.update({'font.size': 5})
fig, ax = plt.subplots(1,1,figsize=(3.5,2.5))

ax.plot(X, ration_f(Px, Qx, X), label=label_f, lw=1.)

ax.set_title('Rational function with singularities')
ax.set_xlabel('x'); ax.set_ylim(-15, 15)
ax.grid(color='r', linestyle=':', linewidth=0.3)
ax.axvline(x=0, c="k", lw=0.4); ax.axhline(y=0, c="k", lw=0.4)

ax.scatter([-3,1,3],[0,0,0], c='r', s=8, label="Real roots of Qx")
xrP = [0]*len(roots_r_P)
ax.scatter(roots_r_P, xrP, c='g', s=8, label="Real roots of Px")
ax.legend(loc='upper left')

plt.savefig('images/ration_f4.png', dpi=500); plt.show()
```

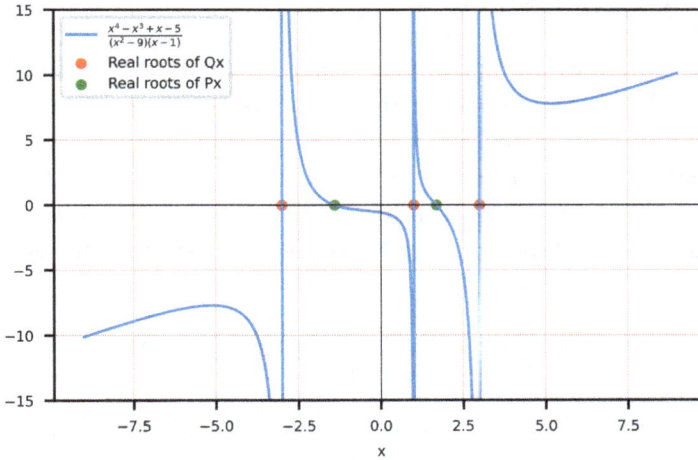

Figure 4.27. A rational function with three singularities (red dots) at the roots of $Q(x)$ and two roots (green dots) that is the roots of $P(x)$.

This function diverges when x approaches ∞ or $-\infty$, as shown in Fig. 4.27 and confirmed by the following code:

```
1  print(f'Limit at left:{limit(ration_f(Px, Qx, sx), sx, -oo)}')
2  print(f'Limit at right:{limit(ration_f(Px, Qx, sx), sx, oo)}')
```

```
Limit at left:-oo
Limit at right:oo
```

```
1  continuous_domain(ration_f(Px, Qx, sx), sx, S.Reals)
```

$$(-\infty, -3.0) \cup (-3.0, 1.0) \cup (1.0, 3.0) \cup (3.0, \infty)$$

This rational function is found continuous in domain $\mathbb{R}$ excluding those singularity points, which is the same as the case we studied earlier. This is because we did not change the denominator polynomial that is the one controlling the singular points.

Rational functions can be used for the interpolation and approximation of functions in computational methods. The following section describes an application as activation function that can be used in machine learning models.

4.15.5 *Rational activation function*

Rational functions are designed as a type of activation function [2] for neural networks. It is equipped with parameter α that can be used to control the peak values of the derivative of the function. α may be treated as a trainable parameter in a machine learning model, so that the derivative value can also be trained. It has also a tunable function $Q(z)$. Therefore, it is a family of activation functions. The general form is defined as follows:

$$\phi(z) = \frac{Q(z)}{\alpha + |Q(z)|}; \quad z \in (-\infty, \infty) \tag{4.38}$$

where α is a constant in $\mathbb{R}$ and $Q(z)$ is an arbitrary function as long as it is (strictly) monotonically increasing with $z \in [0, \infty)$ and monotonically increasing with $z \in (-\infty, 0)$. Thus, the codomain of $\phi(z)$ is in $(-1, 1)$.

4.15.6 *Major properties of rational functions*

The major properties are as follows:

1. The roots of the numerator polynomial are the roots of the rational function.
2. The roots of the denominator polynomial are the singular points of the rational function.
3. If the order of the numerator polynomial is higher than that of the denominator polynomial, the rational function will diverge at the infinities. Otherwise, it will converge.

4.15.7 *Example of rational activation function*

Let, for example, $Q(z) = z$, we have the simplest rational activation function:

$$\phi(z) = \frac{z}{\alpha + |z|} \tag{4.39}$$

If we want it being positive, $\phi(z) \in (0,1)$, we simply shift it by 1.0 and then a scaling by 0.5, which gives

$$\phi(z) = 0.5 \left(1 + \frac{z}{\alpha + |z|} \right) \qquad (4.40)$$

The following Python code computes and plots this activation function:

```python
# Rational activation function in (-1, 1)
def raf(z, alfa):
    f = [zi/(alfa-zi) if zi < 0.0 else zi/(alfa+zi) for zi in z]
    return f

alfa = 2.0;  x = np.arange(-10., 10., .1)       #x -> z
plt.figure(figsize = (4., 1.5), dpi=150.)

plt.plot(x, raf(x,alfa))

plt.xlabel('z')              # in ML, the arguments is denoted as z
plt.ylabel('Function value')
plt.title('Rational activation function, in (-1, 1)')
plt.grid(color='r', linestyle=':', linewidth=0.3)
plt.axvline(x=0, c="k", lw=.5); plt.axhline(y=0, c="k", lw=.5)
plt.savefig('images/raf.png',dpi=500,bbox_inches='tight')
plt.show()
```

This function approaches 1 and -1, respectively, when x approaches ∞ or $-\infty$, as shown in Fig. 4.28, which is a typical behavior of an activation function used in machine learning models [2].

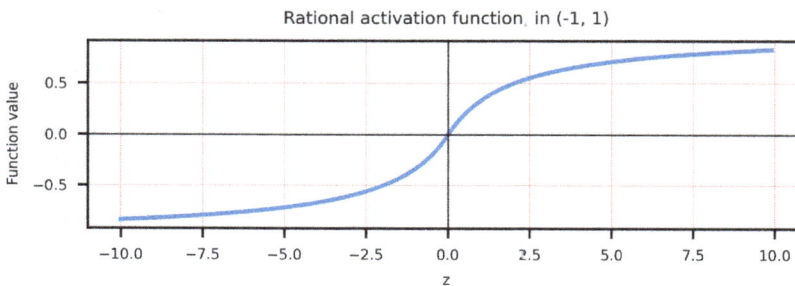

Figure 4.28. A rational function used as activation function for machine learning models.

4.16 Exponential functions

4.16.1 *Definition*

The exponential function is another widely used function in computational methods in different fields to model a variety of phenomena. It is studied intensively in mathematics because of its importance, and many useful properties are known. The following is one of the most well-known formula for computing the compound interest of an investment. Assume that one invests M (constant) dollars in a bank for x years. The bank offers an annual interest rate of r, the return becomes

$$R(1) = M(1+r) \qquad \text{at the end of the first year}$$
$$R(2) = (M(1+r))(1+r) = M(1+r)^2 \quad \text{at the end of the second year}$$
$$\vdots$$
$$R(x) = M(1+r)^x \qquad \text{at the end of the } x\text{th year} \tag{4.41}$$

We now have a function $R(x)$ with the independent variable x raised as a power. This is a typical exponential function. The $(1+r)$ is called the base. A similar formula applies to the population growth (or decay) of species in nature and growth of many field variables in engineering. In other words, any phenomenon involves growth and decay, and the exponential function can have a role.

It is also useful in mathematical operations, such as controlling the convergence of functions at the infinities, using its exponentially fast decaying feature.

In general, an exponential function is denoted as

$$f(x) = b^x \tag{4.42}$$

where b is called base, which is a given constant. The independent variable x raised up as the exponent and thus the name of this function. In general, $x \in \mathbb{C}$. The base b is often confined in $\mathbb{R}$, and it can also be in theory in $\mathbb{C}$.

Most widely used bases are 2, 10, and e known as Euler's number $e \approx 2.718$. The most often used is e^x called the natural exponential function. It is so often that when mentioning exponential function, most likely it refers to

$$f(x) = e^x \quad \text{or} \quad f(x) = \exp(x) \tag{4.43}$$

4.16.2 *Python examples*

Let us plot some exponential functions:

```
1  plt.rcParams.update({'font.size': 10})
2  bases = [1/10, 1/np.e, 1/2, 2, np.e, 10]
3  labels= ['$0.1^x$','$0.5^x$','$e^{-x}$','$2^x$','$e^x$','$10^x$']
4
5  x = np.linspace(-.8,.8,200)        # create values of arguments, x
6  for i in range(len(bases)):
7      plt.plot(x, bases[i]**x, label =labels[i])
8
9  plt.grid(color='r', linestyle=':', linewidth=0.3)
10 plt.axvline(x=0, c="k", lw=.5); plt.axhline(y=0, c="k", lw =.5)
11 plt.scatter(0, 1, c='r', s=25, label='Intersection point')
12 plt.xlabel('x')
13 plt.legend()
14
15 plt.savefig('images/f-exp_x_.png', dpi=500)
16 plt.show();
```

Figure 4.29 plots widely used exponential functions. These functions are all continuous, vary monotonically, and always positive for real argument x. They grow extremely fast with x when base is larger than 1. This is known as exponential growth. They decay with x when base is smaller than 1. The properties of the exponential function are listed as follows:

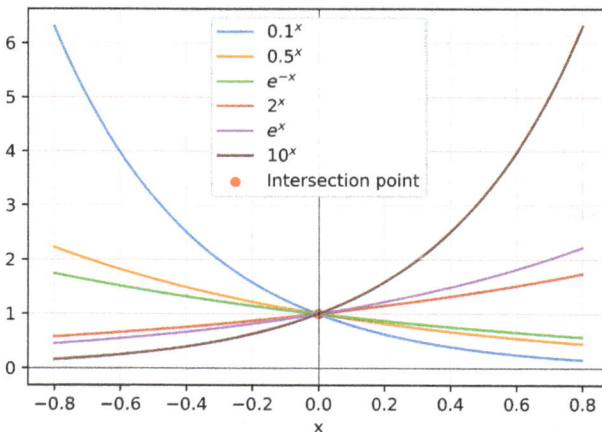

Figure 4.29. Exponential functions: natural base, base 2, and base 10. These functions are monotonic, positive, and grow or decay extremely fast.

4.16.3 *Major properties of exponential functions*

In the following discussions, we focus on e^x because it is the most widely used one and is largely representative.

The major properties are as follows:

- Exponentiation identity:

$$e^{x+y} = e^x e^y \quad \text{for all } x, y \in \mathbb{C} \tag{4.44}$$

This is useful because it allows an addition operation $(x + y)$ to replace a multiplication $(e^x \, e^y)$.

- Exponential functions with another base b (positive and real) can be given by

$$b^x = e^{x \ln b} \tag{4.45}$$

where $\ln()$ is the logarithm to base e, which will be discussed in the following section. This relation is useful to convert exponential functions with different basis. The change in base is compensated by a factor to the exponent.

- Both x and b in Eq. (4.45) can be extended to complex.
- The exponential and trigonometric functions are related via Euler's formula given in Eq. (4.25).

4.16.4 *Sympy examples*

Let us confirm these properties using Sympy:

```
1  x, a1, b1, a2, b2 = sp.symbols("x, a1, b1, a2, b2",real=True)
2  c1 = a1 + sp.I*b1
3  c2 = a2 + sp.I*b2
4  sp.exp(c1 + c2).expand()==(sp.exp(c1)*sp.exp(c2)).expand()
```

True

```
1  b = sp.symbols("b", nonnegative=True)
2  sp.E**(x*sp.log(b, sp.E)).subs({x:3,b:2}).evalf()
```

8.0

```
1  (b**x).subs({x:3,b:2}).evalf()
```

8.0

```
1  (sp.E**2)**3
```

$$e^6$$

4.16.5 *Numpy examples*

Let take a look a few Numpy examples:

```
1  bases = np.array([2, 2j, np.e, np.e*1j])          # base can be complex
2  x = 5.0
3  for i in range(len(bases)):
4      print(f'{bases[i]:.4f}**{x:.1f} = {(bases[i]**x):.4f}')
```

```
2.0000+0.0000j**5.0 = 32.0000+0.0000j
0.0000+2.0000j**5.0 = 0.0000+32.0000j
2.7183+0.0000j**5.0 = 148.4132+0.0000j
0.0000+2.7183j**5.0 = 0.0000+148.4132j
```

```
1  bases = np.array([2, 1j, np.e, np.e*1j])          # base can be complex
2  x = 1j                                            # x can also be complex
3
4  for i in range(len(bases)):
5      print(f'{bases[i]:.4f}**{x:.1f} = {(bases[i]**x):.4f}')
```

```
2.0000+0.0000j**0.0+1.0j = 0.7692+0.6390j
0.0000+1.0000j**0.0+1.0j = 0.2079+0.0000j
2.7183+0.0000j**0.0+1.0j = 0.5403+0.8415j
0.0000+2.7183j**0.0+1.0j = 0.1123+0.1749j
```

One may note that $i^i \approx 0.2079$, which was found using Euler's formula given in Eq. (3.11).

4.16.6 *Useful functions written in e^x*

Many important functions are formed using the exponential function. A few most often-used ones in computational methods are listed as follows:

- hyperbolic sine that uses the odd part of the exponential function:

$$\sinh x = \frac{e^x - e^{-x}}{2} = \frac{e^{2x} - 1}{2e^x} \tag{4.46}$$

- hyperbolic cosine that uses the even part of the exponential function:

$$\cosh x = \frac{e^x + e^{-x}}{2} = \frac{e^{2x} + 1}{2e^x} \tag{4.47}$$

- hyperbolic tangent that is the ratio of sinh and cosh:

$$\tanh x = \frac{\sinh x}{\cosh x} = \frac{e^x - e^{-x}}{e^x + e^{-x}} = \frac{e^{2x} - 1}{e^{2x} + 1} \qquad (4.48)$$

The sinh(), cosh(), tanh() are often used as the basis function to form solutions for mechanics problem together with the polynomials and exponential functions. The tanh is one of the often-used activation function in machine learning models [2]. The following code plots sinh(), cosh() together with exp():

```python
plt.rcParams.update({'font.size': 10})
x = np.linspace(-1.5,1.5,200)        # create values of arguments, x

plt.plot(x, np.exp(x), label ='exp(x)')
plt.plot(x, np.sinh(x), label ='Hyperbolic sine')
plt.plot(x, np.cosh(x), label ='Hyperbolic cosine')
plt.plot(x, np.tanh(x), label ='Hyperbolic tangent')

plt.grid(color='r', linestyle=':', linewidth=0.5)
plt.axvline(x=0, c="k", lw=.5); plt.axhline(y=0, c="k", lw =.5)
plt.xlabel('x')
plt.legend()

plt.savefig('images/Hyperbolic_x_.png', dpi=500)
plt.show();
```

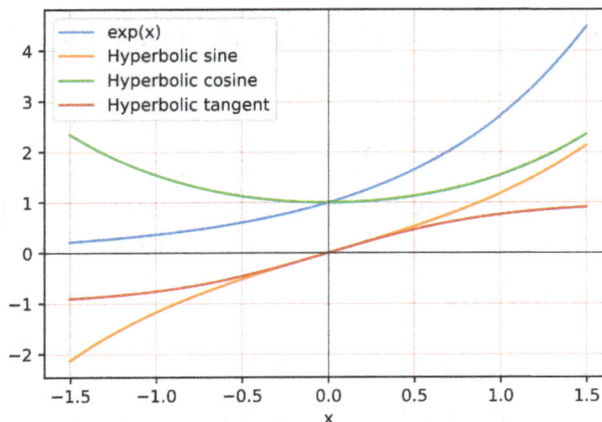

Figure 4.30. Hyperbolic functions: rooted at the natural exponential function; hyperbolic cosine: symmetric, positive; hyperbolic sine tangent: anti-symmetric, monotonical.

Figure 4.30 plots the hyperbolic functions that are made of the natural exponential function.

4.16.7 *Example: Decaying oscillatory functions*

In studying elastic waves in solids including composite laminates, we often encounter functions in the frequency domain [6]. For a given frequency, a typical such a function can be written in the following function of time:

$$f(t) = e^{-\eta t}e^{-i\omega t} = e^{-\eta t}(\cos\omega t - i\sin\omega t) \tag{4.49}$$

where η is a real positive number depending on the material property, ω is the given angular frequency, and t is the time. Note that the codomain of the function is complex, although its domain (for time) is real and positive. The first term controls the decay in time, and the second term is sinusoidal, resulting in a decaying oscillatory function.

We can write the following code to plot this function with one curve for the real part and another for the imaginary part of the function:

```
 1  eta = 2.0; omega = 9.*np.pi
 2  t = np.linspace(0, 1.0, 200)                              # time domain
 3
 4  plt.rcParams.update({'font.size': 10})
 5  plt.plot(t, np.exp(-eta*t)*np.cos(omega*t), label ='Real part')
 6  plt.plot(t,-np.exp(-eta*t)*np.sin(omega*t), label ='Imaginary part')
 7
 8  plt.xlabel('time')
 9  plt.grid(color='r', linestyle=':', linewidth=0.5)
10  plt.axvline(x=0, c="k", lw=.5); plt.axhline(y=0, c="k", lw =.5)
11  plt.legend()
12
13  plt.savefig('images/decayingWavelet.png', dpi=500)
14  plt.show();
```

The function varies with time sinusoidally, but its amplitude decays with time, as shown in Fig. 4.31. Such an exponential decay ensures the convergence with time, and η relates to the dumping coefficient of the material (and

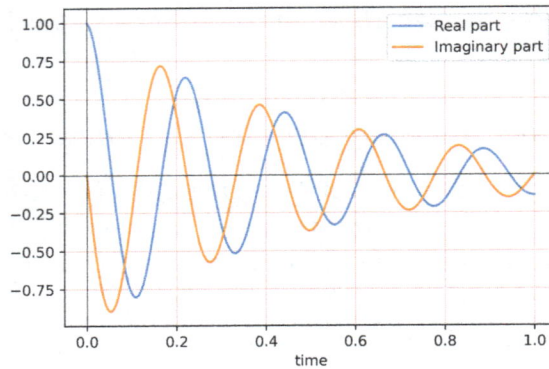

Figure 4.31. Real and imaginary parts of a waveform decaying with time.

hence always positive). This is a typical feature of a function in nature and engineering systems when analyzed in a given frequency. The combination of exponential and trigonometric functions provides a good representation of the behavior [6]. This type of behavior is also observed in the vibration of engineering structures.

4.17 Logarithm functions

4.17.1 *Definition*

A logarithm function is the inverse function of the corresponding exponential functions discussed above. Its domain $\mathbb{X}$ is $(0, \infty)$ and codomain $\mathbb{Y}$ is $(-\infty, \infty)$, all of which can be in $\mathbb{R}$. It is conventionally denoted as

$$y = f(x) = \log_b x \tag{4.50}$$

where b is the base and is in $(0, \infty) \in \mathbb{R}$ but $b \neq 1$. When $b = 1$, x and y are not related at all. This logarithm function is the inverse of the exponential function $x = b^y$:

$$\log_b(x) = \log_b(b^y) = y \tag{4.51}$$

For example, $8 = 2^3$ in exponential form, the logarithm base 2 of 8 is 3, or $\log_2 8 = 3$, as shown by the blue circled point in Fig. 4.32.

4.17.2 *Change of bases*

When base b gets bigger, the curve becomes closer to the x-axis. Figure 4.33 plots logarithm functions with three often-used bases: 2, e, and 10.

Let us check the continuous domain of the natural logarithmic function defined in the real space.

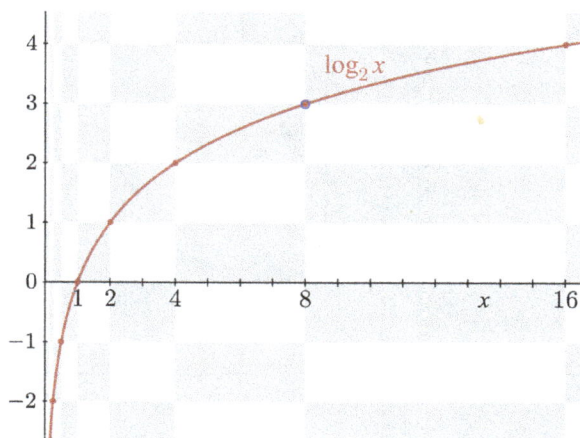

Figure 4.32. The logarithm function with base 2. It passes through points $(2, 1)$, $(4, 2)$, and $(8, 3)$ for $\log 2(8) = 3$ and $2^3 = 8$. The curve can get arbitrarily close to the y-axis, but will never meet it. Modified based on the image from Wikipedia Commons originally made by Jacob Rus under the CC BY-SA 4.0 license.

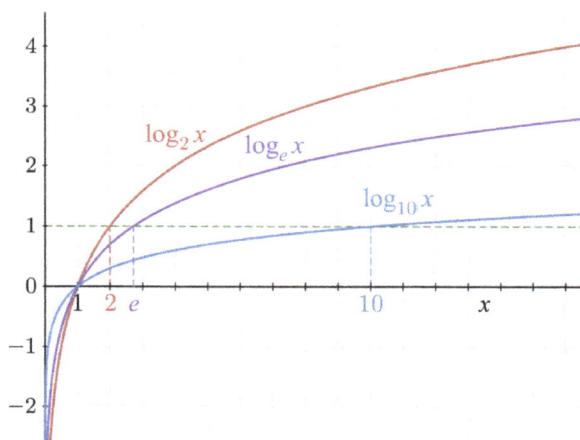

Figure 4.33. Logarithm functions with three commonly used bases. The special points $\log_b b = 1$ are at the dotted horizontal line. All three curves intersect in $log_b 1 = 0$. Image from Wikipedia page made by Richard F. Lyon under the CC BY-SA 3.0 license.

```
1  from sympy import limit, oo, Symbol, S
2  from sympy.calculus.util import continuous_domain
3
4  sx = Symbol('sx')          # sx: symbolic x
5  continuous_domain(sp.log(sx), sx, S.Reals)
```

$(0, \infty)$

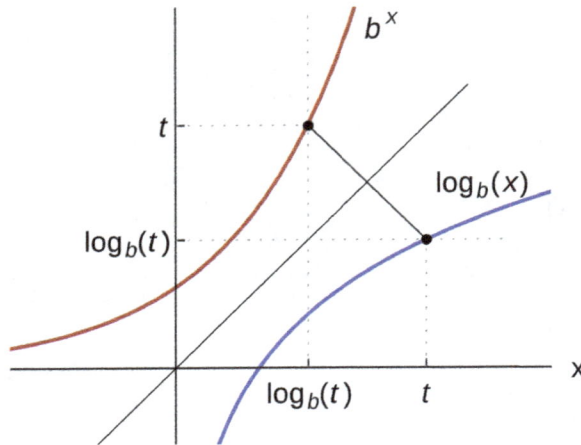

Figure 4.34. The logarithm function is the inverse or mirror image of an exponential function. The "mirror line" is at 45° angle. The image is from Wikimedia Commons made by Stpasha under the CC0 1.0 Universal license.

4.17.3 *Mutual inverse with the exponential function*

The logarithm function extracts the exponent of a large argument using a base. It is also a mirror image of an exponential function, as shown in Fig. 4.34.

4.17.4 *Major properties*

In computational methods, $\log_{10}()$ is called the decimal or common logarithm, $\log_2()$ is the binary logarithm, and $\log_e()$ is called the natural logarithm, which is often written as $\ln()$.

Logarithm functions have the following useful properties:

- **Product:** The logarithm of a product of two arguments is the sum of the logarithms of these two arguments:

$$\log_b(xy) = \log_b x + \log_b y \qquad (4.52)$$

This is useful because it converts multiplication to addition.

- **Division:** The logarithm of an argument divided by another is the difference of the logarithms of these two arguments:

$$\log_b\left(\frac{x}{y}\right) = \log_b x - \log_b y \qquad (4.53)$$

This converts devision to subtraction.

- **Power:** The logarithm of an argument with a power $p \in \mathbb{R}$ is the logarithm of the argument multiplied by the power:

$$\log_b(x^p) = p \log_b x \qquad (4.54)$$

This is useful because it converts power to multiplication.

- **Change of base:** $\log_b x$ can be computed from the logarithms of x and b with respect to an arbitrary common base k:

$$\log_b x = \frac{\log_k x}{\log_k b} \qquad (4.55)$$

4.17.5 *Python examples of functions*

The following are some examples of operations using Python:

```python
# a few special values:
print(np.log2 ([1,    2,    2**2, 1000, 0.0000001]))
print(np.log  ([1,np.e,np.e**2, 1000, 0.0000001]))
print(np.log10([1,    10,   10**2, 1000, 0.0000001]))
```

```
[ 0.           1.           2.           9.96578428  -23.25349666]
[ 0.           1.           2.           6.90775528  -16.11809565]
[ 0.  1.  2.  3. -7.]
```

Let us plot $\log_5(x)$ and $\log_{1/5}(x)$ using the following code to observe interesting effect of the base for logarithm:

```python
plt.rcParams.update({'font.size': 5})
fig, ax = plt.subplots(1,1,figsize=(3,2))
x = np.arange(0.001, 5, .01)

def logb(b,x):                          # compute log_b using log_e
    return np.log(x)/np.log(b)          # log is log_e in numpy

ax.plot(x,  logb(5,x), lw=.8, label="$y=log_5(x)$")
ax.plot(x,logb(1/5,x), lw=.8, label="$y=log_{1/5}(x)$")

ax.set_xlabel('x'); ax.set_ylabel("$y(x)$")
ax.grid(color='r', linestyle=':', linewidth=0.3)
plt.axvline(x=0, c="k", lw=.2); plt.axhline(y=0, c="k", lw =.2)
ax.legend()

plt.savefig('images/log2nd5.png', dpi=500)
plt.show()
```

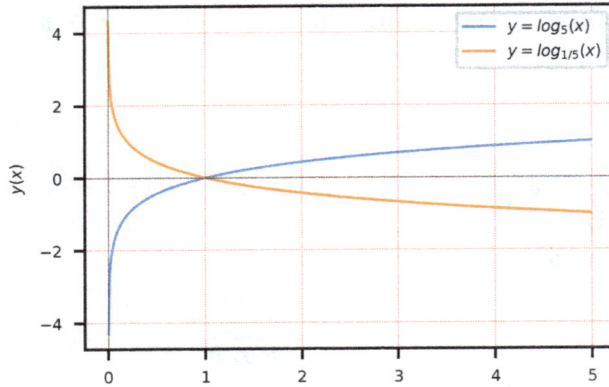

Figure 4.35.   Logarithm functions with bases: $b = 5$ and $b = 1/5$.

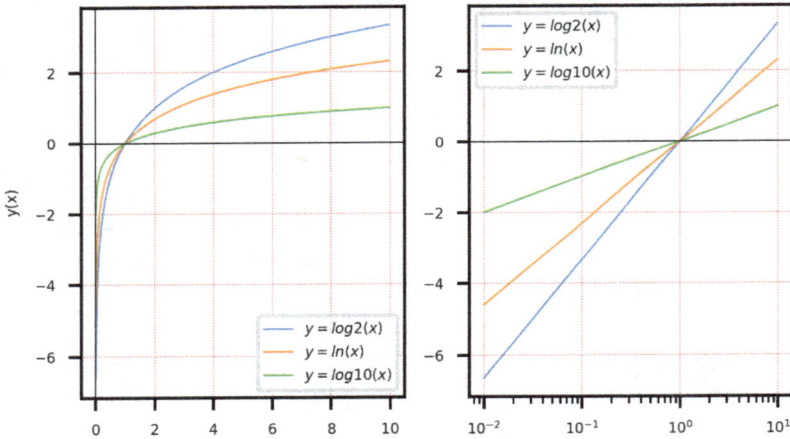

Figure 4.36.   Three logarithm functions.

4.17.6 *Logarithmic scaling of functions*

Since the argument of a function can also be another function, logarithmic function is thus useful to drastically condense the variable or another function value that is larger than 1, or magnify the value that is smaller than 1. This is done without loss of continuity and monotonicity of the argument functions. It has widespread application, including machine learning [2].

One of the most widespread use of logarithm is for plotting curves in the so-called logarithmic axis. This can often make the plot easy to read. For example, when logarithm functions are plotted with x-axis set to be

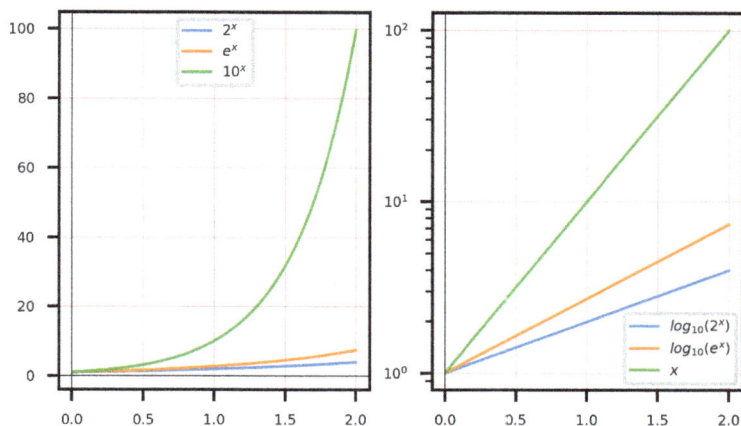

Figure 4.37. Three exponential functions.

logarithmic, these curves become simple straight lines, as produced using the following code:

```
1  plt.rcParams.update({'font.size': 4})
2  fig, ax = plt.subplots(1,2,figsize=(3.5,2.))
3  xlabel = ['x', 'log(x)']
4  x = np.arange(0.01, 10., .01)
5
6  for i in range(len(ax)):
7      ax[i].plot(x,  np.log2(x), lw=.5, label="$y=log2(x)$")
8      ax[i].plot(x,   np.log(x), lw=.5, label="$y=ln(x)$")
9      ax[i].plot(x, np.log10(x), lw=.5, label="$y=log10(x)$")
10
11     ax[i].set_xlabel(xlabel[i])
12     ax[i].grid(color='r', linestyle=':', linewidth=0.3)
13     ax[i].axvline(x=0, c="k", lw=.3); ax[i].axhline(y=0, c="k", lw =.3)
14     ax[i].legend()
15
16 ax[0].set_ylabel('y(x)')
17 ax[1].set_xscale('log')
18 plt.savefig('images/ThreeLogs.png', dpi=500)
19 plt.show()
```

4.17.7 *Logarithmic scaling for exponential functions*

On the other hand, when exponential functions are plotted with y-axis set to be logarithmic, these curves also become simple straight lines, as produced using the following code. This is useful in examining the convergence rate of a computational method, where the slope of the straight line is the convergence rate [1].

```
1  plt.rcParams.update({'font.size': 4})
2  fig, ax = plt.subplots(1,2,figsize=(3.5,2.))
3  x = np.arange(0.01, 2., .001)
4
5  #ylabel = ['y(x)', 'log(y(x))']
6  labels = [("$2^x$", "$e^x$", "$10^x$"),
7            ("$log_{10}(2^x)$", "$log_{10}(e^x)$", "$x$")]
8
9  for i in range(len(ax)):
10     ax[i].plot(x,    2.**x, lw=.8, label=labels[i][0])
11     ax[i].plot(x,np.exp(x), lw=.8, label=labels[i][1])
12     ax[i].plot(x,   10.**x, lw=.8, label=labels[i][2])
13
14     ax[i].set_xlabel('x')
15     #ax[i].set_ylabel(ylabel[i])
16     ax[i].grid(color='r', linestyle=':', linewidth=0.2)
17     ax[i].axvline(x=0,c="k",lw=.3); ax[i].axhline(y=0,c="k",lw =.3)
18     ax[i].legend()
19
20  ax[1].set_yscale('log')
21  plt.savefig('images/Threeexps.png', dpi=500)
22  plt.show()
```

4.17.8 *In complex domain*

Logarithmic function can also have complex independent variables, including negative real values. In this case, the output is in general also complex-valued. Following is the Numpy code for demonstration. Since we used zero as an input, we turn off the warning:

```
1  np.seterr(divide='ignore')
2  bases = np.array([0, 1, -2, 2j, np.e, np.e*1j])    # base can be complex
3
4  for i in range(len(bases)):
5      print(f'log({bases[i]:.4f}) = {np.log(bases[i]):.4f}')
```

```
log(0.0000+0.0000j) = -inf+0.0000j
log(1.0000+0.0000j) = 0.0000+0.0000j
log(-2.0000+0.0000j) = 0.6931+3.1416j
log(0.0000+2.0000j) = 0.6931+1.5708j
log(2.7183+0.0000j) = 1.0000+0.0000j
log(0.0000+2.7183j) = 1.0000+1.5708j
```

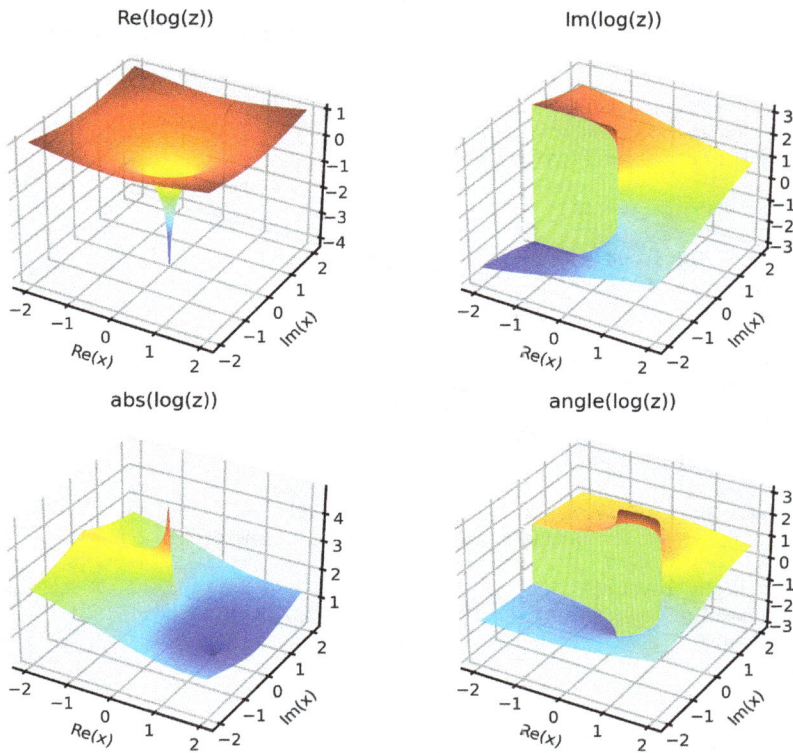

Figure 4.38. Logarithmic function in complex domain.

```
1  np.seterr(divide='warn')          # turn back on the standard warning
```

```
{'divide': 'ignore', 'over': 'warn', 'under': 'ignore', 'invalid': 'warn'}
```

```
1  xR = np.linspace(-2 ,2, 200)      # range cn the real axis
2  xI = np.linspace(-2.,2, 200)      # range for the imaginary axis
3  lxR, lxI = len(xR), len(xI)
4
5  XR, XI = np.meshgrid(xR, xI)
6  logz = np.log(XR + XI*1j)         # complex
```

```
7  Z = (      logz.real,   logz.imag,   np.abs(logz), np.angle(logz))
8  z_labels=['Re(log(z))','Im(log(z))','abs(log(z))','angle(log(z))']
9
10 plt.rcParams.update({'font.size': 6})
11 plt.rcParams["figure.autolayout"] = True
12 fig_s = plt.figure(figsize=(6,4))
13
14 for i in range(len(Z)):
15
16     ax = fig_s.add_subplot(2,2,i+1, projection='3d')
17     ax.set_xlabel('Re(x)', labelpad=-11)
18     ax.set_ylabel('Im(x)', labelpad=-11)
19
20     ax.tick_params(axis='x', pad=-5)
21     ax.tick_params(axis='y', pad=-5)
22     ax.tick_params(axis='z', pad=-5)
23     #ax.set_zlabel(z_labels[i], rotation=0)
24     plt.title(z_labels[i], pad=-25)
25
26     ax.plot_surface(XR,XI,Z[i],color='b',rstride=1,cstride=1,
27                     shade=False,cmap="jet", linewidth=1)
28
29 fig_s.tight_layout()  # otherwise right y-label is clipped
30 plt.savefig('images/logz_complex.png',dpi=500,bbox_inches='tight')
31 plt.show()
```

The surfaces of the complex logarithmic function are plotted in Fig. 4.38. The singular point is at the origin where both real and imaginary parts are zero.

The following generates contour plots of the logarithm function in complex domain:

```
1  fig_s = plt.figure(figsize=(6,6))
2
3  for i in range(len(Z)):
4      ax = fig_s.add_subplot(2,2,i+1)
5      ax.set_xlabel('Re(x)'); ax.set_ylabel('Im(x)')
6      ax.axvline(x=0,c="k",lw=.3); ax.axhline(y=0,c="k",lw =.3)
7      plt.title(z_labels[i])
8
9      ax.contour(XR,XI,Z[i], levels=40, linewidths=0.8)
10
11 fig_s.tight_layout()  # otherwise right y-label is clipped
12 plt.savefig('images/logz_complex_c.png',dpi=500,bbox_inches='tight')
13 plt.show()
```

The contours of the complex logarithmic function is plotted in Fig. 4.39.

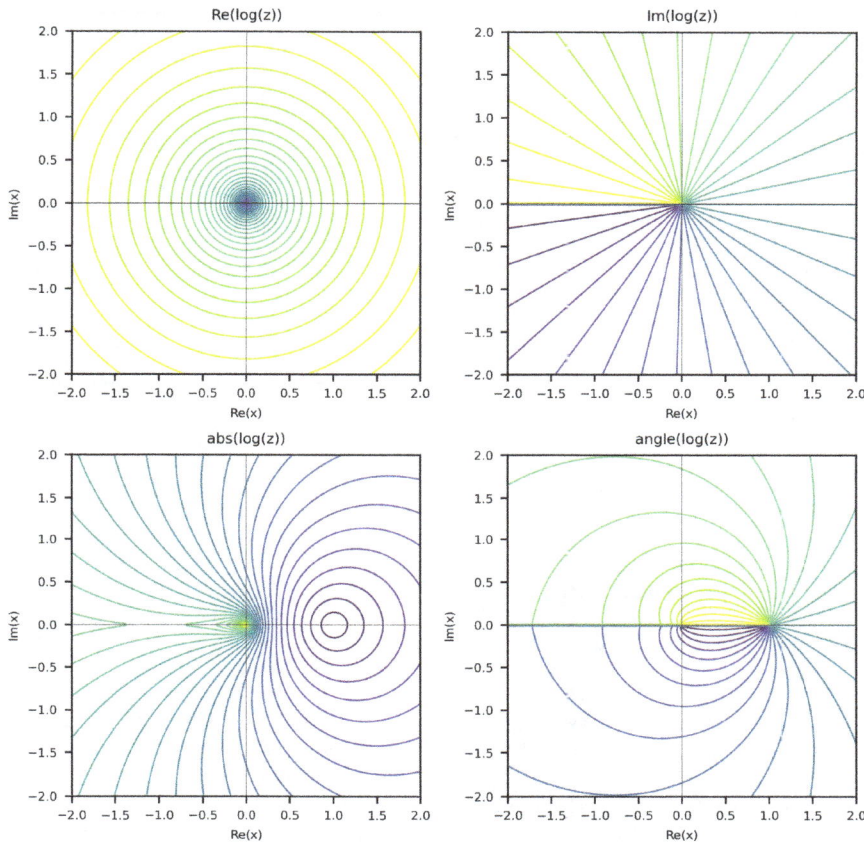

Figure 4.39. Contour of a logarithmic function in complex domain.

4.18 The x^x function

As we defined in the beginning of this chapter, a function is essentially a mathematic device that maps a set of variables (domain) to another set of variables (codomain). Therefore, there can be infinite types of functions. One can create one if one can imagine one. Just to support the "imagination is the limit" argument, this section discusses a very strange function:

$$f(x) = x^x \tag{4.56}$$

The independent variable x is the base and also the exponent! Clearly, if the domain $\mathbb{X}$ is $[0, \infty)$, the codomain $\mathbb{Y}$ is also $[0, \infty)$. Here, we note that $0^0 = 0$. It is smooth but varies extremely fast in the codomain, with nothing particularly strange, as shown in the plot produced using the following code:

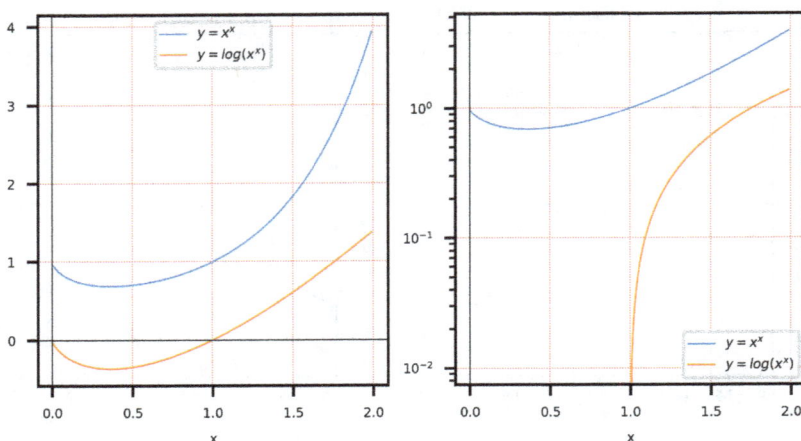

Figure 4.40. Two interesting functions when $x > 0$.

```python
1  plt.rcParams.update({'font.size': 4})
2  fig, ax = plt.subplots(1,2,figsize=(3.5,2.))
3  xlabel = ['x', 'x']
4  x = np.arange(0.01, 2., .01)
5
6  for i in range(len(ax)):
7      ax[i].plot(x,  x**x, lw=.5, label="$y=x^x$")
8      ax[i].plot(x, np.log(x**x), lw=.5, label="$y=log(x^x)$")
9
10     ax[i].set_xlabel(xlabel[i])
11     ax[i].grid(color='r', linestyle=':', linewidth=0.3)
12     ax[i].axvline(x=0, c="k", lw=.3); ax[i].axhline(y=0, c="k", lw =.3)
13     ax[i].legend()
14
15 ax[1].set_yscale('log')
16 plt.savefig('images/xxLogxx.png', dpi=500)
17 plt.show()
```

It is seen from Fig. 4.40 that even the $\log(x^x)$ varies very fast when x approaches 1.

When x is negative, the codomain of this function requires more detailed analysis. Using Eq. (3.12), for $x < 0$, we have

$$
\begin{aligned}
f(x) = x^x &= |x|^x e^{i(2k\pi+\pi)x} \\
&= |x|^x (\cos[(2k\pi + \pi)x] + i\sin[(2k\pi + \pi)x])
\end{aligned}
\tag{4.57}
$$

Considering the primary domain of $(-\pi, \pi]$ for the phase angle, we set $k = 0$, which gives

$$
f(x) = x^x = |x|^x e^{i\pi x} = |x|^x (\cos(\pi x) + i\sin(\pi x))
\tag{4.58}
$$

It is clear that the function value switches between real and complex numbers. The codomain of this function has only discrete points ($x = 0$, $-1, -2, \ldots,$) in the real space, but continuously oscillating in the complex space.

If the absolute value of k grows in Eq. (4.57), the switching gets faster. The discrete points in the real space get denser. In the complex space, the function oscillate more frequently and continuously.

Readers may take a look at the online video at https://www.youtube com/watch?v=6HYZWVYv0WY (in Chinese) for a detailed analysis.

Here, let us plot this interesting function in the complex space:

```python
xR = np.linspace(-1*np.pi, 1*np.pi, 200)          # along real axis
xI = np.linspace(-2*np.pi, 2*np.pi, 400)          # along imaginary axis
lxR, lxI = len(xR), len(xI)

XR, XI = np.meshgrid(xR, xI)
logz = (XR + XI*1j)**(XR + XI*1j)          # x^x, in complex
Z = (     logz.real,  logz.imag, np.abs(logz), np.angle(logz))
z_labels=['$Re(x^x)$','$Im(x^x)$','$abs(x^x)$','$angle(x^x)$']

plt.rcParams.update({'font.size': 6})
plt.rcParams["figure.autolayout"] = True
fig_s = plt.figure(figsize=(6,4))

for i in range(len(Z)):

    ax = fig_s.add_subplot(2,2,i+1, projection='3d')
    ax.set_xlabel('Im(x)', labelpad=-11)
    ax.set_ylabel('Re(x)', labelpad=-11)

    ax.tick_params(axis='x', pad=-5)
    ax.tick_params(axis='y', pad=-5)
    ax.tick_params(axis='z', pad=-5)
    #ax.set_zlabel(z_labels[i], rotation=0)
    plt.title(z_labels[i], pad=-25)

    ax.plot_surface(XI,XR,Z[i],color='b',rstride=1,cstride=1,
                    shade=False,cmap="jet", linewidth=1)

fig_s.tight_layout()  # otherwise right y-label is clipped
plt.savefig('images/xx_complex.png',dpi=500,bbox_inches='tight')
plt.show()
```

It is seen from Fig. 4.41 that both the real and imaginary parts of x^x varies very fast at $Re(x) = 0$.

Let us plot its contour plots:

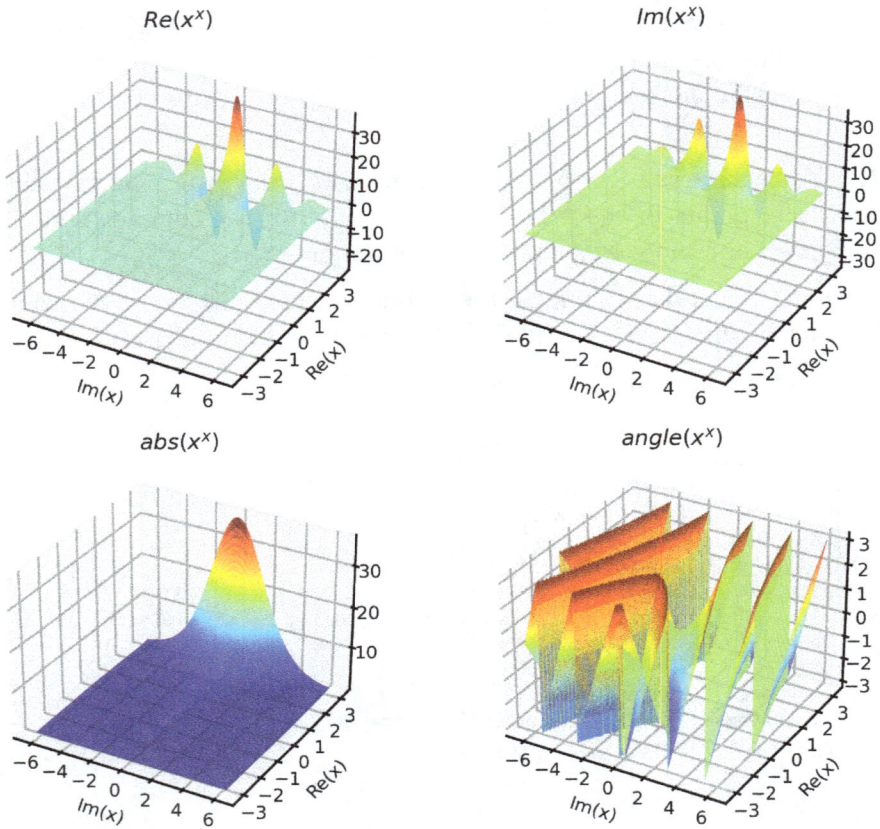

Figure 4.41.   Surfaces of function x^x in complex domain.

```
1  fig_s = plt.figure(figsize=(6,6))
2
3  for i in range(len(Z)):
4
5      ax = fig_s.add_subplot(2,2,i+1)
6      ax.set_xlabel('Im(x)'); ax.set_ylabel('Re(x)')
7      ax.axvline(x=0,c="k",lw=.3); ax.axhline(y=0,c="k",lw =.3)
8      plt.title(z_labels[i])
9
10     ax.contour(XR,XI,Z[i], levels=40, linewidths=0.8)
11
12 fig_s.tight_layout()  # otherwise right y-label is clipped
13 plt.savefig('images/xx_complex_c.png',dpi=500,bbox_inches='tight')
14 plt.show()
```

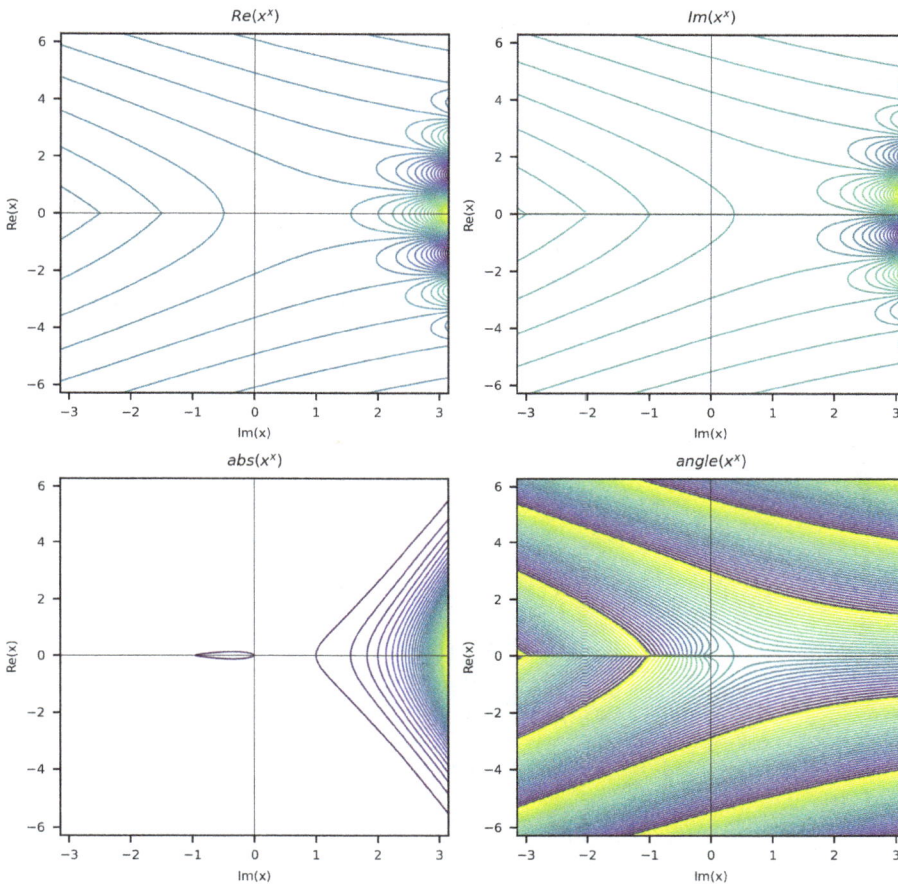

Figure 4.42. Contours of function x^x in complex domain.

4.19 Composite functions

Functions can be combined to form new function for various purposes. A typical example is the loss function used in a machine learning model, which can be a chain of thousands of functions [6]. Thus, the basic concepts of composite function is of importance, which is briefed as the final topic of this chapter.

4.19.1 *Definition and change of domains*

- Consider two functions $f \colon \mathbb{X} \to \mathbb{Y}$ and $g \colon \mathbb{Y} \to \mathbb{Z}$.
- Then, the codomain of f is the domain of g.
- The composition of these two functions becomes a new function $g(f(x))$ that has a domain $\mathbb{X}$ and a codomain $\mathbb{Z}$.

4.19.2 *Composition in order*

Note that, in general, $g(f(x))$ is different from $f(g(x))$ even if the domain of f is the codomain of g.

For example, let $f(x) = ax^2$ and $g(x) = bx + c$, where a, b, and c are all real constant numbers, and their domain and codomain are the same as $\mathbb{R}$.

We have $g(f(x)) = abx^2 + c$, which is different from $f(g(x)) = a(bx + c)^2$.

Therefore, the composition order of evaluations should be kept during computation.

Let us take a look at some Sympy examples:

```
1  x, y, a, b, c = sp.symbols('x, y, a, b, c ')
2  f = sp.Function('f')
3  g = sp.Function('g')
```

```
1  f = a*x**2
2  g = b*f+c            # a composite function
3  g
```

$$abx^2 + c$$

```
1  g = b*x + c
2  f = a*g**2           # another composite function
3  f
```

$$a(bx + c)^2$$

```
1  f = a*sp.sin(x)
2  g = b*sp.exp(f)+c    # a composite function
3  g
```

$$be^{a\sin(x)} + c$$

```
1  g = b*sp.exp(x)+c
2  f = a*sp.sin(g)      # another composite function
3  f
```

$$a\sin(be^x + c)$$

Let us find the roots of this composite function.

```
1  sp.solve(f,x)
```

$$\left[\log\left(-\frac{c}{b}\right),\ \log\left(\frac{\pi - c}{b}\right)\right]$$

4.20 Extended topics on functions

4.20.1 *Scalar functions in multi-dimensions*

There are many scalar functions that depend on several independent variables, and hence, the domains are multi-dimensional, while the codomain is still one-dimensional. For example, the temperature in room is usually different from one location to another in three-dimensional space. This means that the function of the room temperature depends on the three coordinate variables, say x, y and z. Since the temperature is a number, it is a scalar function of x, y and z. Mechanics problems for solids, structures, and fluids are some such typical problems [5,6].

The simplest way to deal with this type of functions is to isolate one of the independent variables at a time while fixing the others unchanged and then treat the function as a one-dimensional. This simple approach works for simple problems with regular domains. If the problem domain gets irregular, much more careful formulations are needed for effective handling. More systematic and effective means are required, which will be discussed in a future volume of this book series.

4.20.2 *Vector functions*

There are also vector functions, implying that the function itself has more than one component. In this case, its codomain is multi-dimensional and its domain may be one- or multi-dimensional. For example, the displacement in a loaded solid has components in x-, y- and z-directions. Each of these components is a function of one or more independent variables. In addition, the domain can often be irregular. The FEM [5], S-FEM [4] and meshfree methods [1, 7] are typical techniques dealing with this type of functions. We plan to discuss about this type of functions in a collective manner in a future volume of this book series, so that readers can have a better idea about the commonalities and features of many different computational methods.

4.21 Remarks

1. In essence, a function is an object just like a number discussed in detail in the previous chapter. The difference is that function is a device that takes numbers as inputs and can generate many numbers in a defined manner.

It is used to present critical quantities and their behavior of practical problems.

2. Functions have a distributive behavior, domain and codomain, roots, limits, continuity and convergence properties. Plots are the best way to examine these features. This chapter provided a large number of Python codes to compute and plot various functions.

3. Functions can have many types, each of which can also have its reciprocal and/or inverse functions.

4. Just like numbers, functions can also form spaces, in which a special type of function can live. Such a space can be formed (spanned) using a set of basis functions, which will be discussed in the following chapter.

5. Functions have a number of important properties that can be local and global, such as derivatives and integrals, which will be discussed in a future volume of this book series.

References

[1]   Liu GR and Han X, *Computational Inverse Techniques in Nondestructive Evaluation.*, Taylor and Francis Group, New York, 2003.

[2]   Liu GR and Quek SS, *The Finite Element Method: A Practical Course*, Butterworth-Heinemann, 2013.

[3]   Liu GR and Nguyen TT, *Smoothed Finite Element Methods*, Taylor and Francis Group, New York, 2010.

[4]   Liu GR, *Mesh Free Methods: Moving Beyond the Finite Element Method*, Taylor and Francis Group, New York, 2010.

[5]   Liu GR and Zhang G-Y, *Smoothed Point Interpolation Methods: G Space Theory and Weakened Weak Forms*, World Scientific, 2013.

[6]   Liu GR, *Machine Learning with Python: Theory and Applications*, World Scientific, 2023.

[7]   Liu GR and Xi ZC, *Elastic Waves in Anisotropic Laminates*, 2001.

Chapter 5

Basis Functions

Contents

Let us import the following Python modules for later use in this chapter:

```
1  import sys
2  sys.path.append('../grbin/')          # for the use of our own module
3  from commonImports import *           # often used external modules
4  import grcodes as gr
5  importlib.reload(gr)          # reload when changes are made to grcodes
6  #plt.rcParams.update({'font.size': 6})      # control plot font size
7  #np.set_printoptions(precision=4, suppress=True)      # print digits
```

In the previous chapter, we found that a scalar function is a mathematical object quite similar to a number. It essentially maps numbers in a space to numbers in another space. Numbers can form vector spaces that are closed under linear combinations, and hence, new members can be created via linear combinations of basis members.

Along with this argument, we shall also form **vector spaces of functions**, and then new functions can be generated via linear combinations of a special type of functions called **basis functions**. This idea can then further be extended to approximate functions using linear combinations of basis functions. This process for approximating functions is particularly useful because it does not use differential and integral properties of the function to be approximated, which we often do not know.

This chapter first presents the concept of function vector space and then focuses on various types of basis functions that can span such spaces. Features and properties of these basis functions will be discussed in great detail and examined using the developed Python codes. The chapter is written in reference to the documentation on related Python, Numpy, Sympy, and other relevant wiki pages, including Chebyshev polynomials, Lagrange polynomial, Legendre polynomials, Laguerre polynomials, and Hermite polynomials. Discussions with ChatGPT, Bing, and Bard along with Google searches were also very helpful.

5.1 Vector space of functions

The concept of vector space for functions is fundamentally important for function approximations. This is because, to approximate a function, we need to use some known functions in a proper manner. We would like to know the property of the function to be approximated. The definition of vector space for functions is essentially the same as that for vector space for vectors discussed in Section 2.11.2. However, the expression is different. Given in the following is a former definition.

Consider a set of continuous scalar functions denoted as $\mathbb{V}$ over a field $\mathbb{F}$. These functions can be the elementary functions presented in the previous chapter and beyond. Assume that all these functions in $\mathbb{V}$ have a common domain $\mathbb{X}$. For $x \in \mathbb{X}$, we then define the following:

1. Zero function:

$$0(x) = 0 \tag{5.1}$$

2. For any function $f(x) \in \mathbb{V}$, there is a $-f(x) \in \mathbb{V}$, such that $f(x) + -f(x) = 0$.
3. $1f(x) = f(x)$, where 1 is the multiplicative identity in $\mathbb{F}$.
4. The addition of two arbitrary functions, $f(x)$ and $g(x)$ in $\mathbb{V}$, is defined as

$$(f + g)(x) = f(x) + g(x) \tag{5.2}$$

5. The scalar multiplication of a function is defined as

$$(\alpha f)(x) = \alpha f(x) \tag{5.3}$$

where α is a scalar in $\mathbb{F}$.

With the above definitions, the set of continuous scalar functions in $\mathbb{V}$ form a vector space because these **axioms** for a vector space can be satisfied, including containing zero, unit, and negative elements, addition associative and commutative, multiplication distributive over scalar additions and over vector additions, and scalar multiplication associative. Readers may try to verify this in reference to the proof for vector spaces for numbers in Chapter 2. We thus note the following:

> Functions with common domain $\mathbb{X}$ can form vector spaces. These spaces are closed under linear combinations.

To form a function vector space, we use a set of basis functions.

5.2 Conditions for being basis functions

The **condition** for a set of functions that can be used as basis functions is that all these functions in the set have a common domain and must be **linearly independent**, implying that any one of them cannot be expressed as a linear combination of the rest in the set. The linear independence is largely achieved in two ways: (1) by using basis functions, each of which has a different order; (2) by creating basis functions, each of which occupy different local domains. This chapter discusses both these ways.

Based on the properties of vector spaces, any function in the space can be uniquely represented using these basis functions.

There are many types of basis functions, but our discussion will be focused on basis functions that stem from monomials because these are most widely used.

5.3 Monomial basis functions

5.3.1 *Monomials*

The first and simplest set of basis functions is the monomials discussed in detail in the previous chapter. These monomials are the basis functions:

$$1, x^1, x^2, \ldots, x^n \tag{5.4}$$

The domain of monomial basis functions is $(-\infty, \infty)$.

5.3.2 *Proof of monomials as basis functions*

It can be easily shown that these monomials are basis functions for a vector space of polynomial functions of order n.

First, this set of distinct monomials are clearly linearly independent because any of them cannot be expressed as a linear combination of the rest. For example, it is not possible to express x^2 by a linear combination of $1, x^1, x^3, x^4, \ldots, x^n$. The linear independence among these basis functions is achieved by having different orders.

Second, any linear combination of any pair of these monomials, say x^i and x^j with $0 \leq i, j \leq n$,

$$\alpha x^i + \beta x^j \tag{5.5}$$

is a polynomial function with order within n for any scalar α and β in $\mathbb{F}$ (which can be $\mathbb{R}$ or $\mathbb{C}$).

The set of basis functions in Eq. (5.4) is **complete** to degree of n. This means that it contains all these monomials with degrees lower than or equal to n. In other words, there is no missing monomial before x^n in Eq. (5.4). Such a set of basis functions forms a function space of all polynomial functions up to degree n. Note that we always use complete basis functions unless there is a special reason for not doing so. If a order term is missing, these basis functions can still form a space, but that space contains only polynomials without that order.

We also note that the codomain of an approximated function is $(-\infty, \infty)$ if the variable x and all the coefficients for the monomials are real numbers. Otherwise, the codomain will be the complex plane because the complex set is closed on linear and nonlinear operations, as discussed in Chapter 3.

5.3.3 *Inverse monomials*

Inverse monomials can also be used as basis functions, which are expressed as

$$1, \frac{1}{x^1}, \frac{1}{x^2}, \cdots, \frac{1}{x^n} \tag{5.6}$$

The domain of monomial basis functions is $(-\infty, 0) \cup (0, \infty)$. These inverse monomials are also linearly independent. They can be used for approximation functions that converge at infinity, for example, in the so-called infinite elements [3].

Note also that the monomials and inverse monomials can also be used together because they are all linearly independent.

5.3.4 *Python examples: Generation of monomial basis*

We now write some Python code to generate monomial basis functions:

```python
import numpy.polynomial as Poly
i = 8                              # specify the order (degree)
monomial = Poly.Polynomial.basis        # create an object
monomial(i)                          # generate on monomial
```

$x \mapsto$ 0.0+0.0 x+0.0 x^2+0.0 x^3+0.0 x^4+0.0 x^5+0.0 x^6+0.0 x^7+1.0 x^8

```python
#help(Poly.Polynomial.basis)
```

One may generate an arbitrary number of monomial basis functions:

```python
N = 3
[monomial(i, domain=[-oo, oo], window=[-1., 1.]) for i in range(N)]
```

```
[Polynomial([1.], domain=[-oo, oo], dtype=object, window=[-1.,  1.]),
 Polynomial([0., 1.], domain=[-oo, oo], dtype=object, window=[-1.,  1.]),
 Polynomial([0., 0., 1.], domain=[-oo, oo], dtype=object, window=[-1.,  1.])]
```

Any polynomials can be generated by a linear combination of monomials by simply providing the coefficients for each of the monomial bases in the ascending order:

```
1  P3 = Poly.Polynomial([8.,-9.,26.,-24.])    # Generate a poly with coeffs
2  P3
```

$$x \mapsto 8.0 - 9.0\,x + 26.0\,x^2 - 24.0\,x^3$$

This effectively shows that any polynomial can be formed using monomials uniquely. It is in fact the definition of a polynomial.

If monomial basis functions are used to express a function rather than polynomials, it would be an approximation, which will be discussed in the following chapter.

Let us use this P3(x) function obtained above to compute its values at some given points on x-axis:

```
1  xi = np.array([1, 2, 3, 4])
2  P3_4_points = P3(xi)
3  P3_4_points
```

```
array([ 1.000e+00, -9.800e+01, -4.330e+02, -1.148e+03])
```

As expected, we obtained four function values, which will be used for the examination of an example in the following section on Lagrange polynomials.

5.4 Lagrange polynomials

A Lagrange polynomial is formed using monomials in a special way, so that it can be used directly for approximating a function in a domain.

5.4.1 *Definition and formulation*

Consider a domain $[a, b]$ (in the finite element method [3], it is called an element) that is discretized into n intervals with $(n + 1)$ nodes: $[a = x_0, x_1, \ldots, b = x_n]$. The discretization does not have to be uniform, but each node must be distinct. If the values of a function $f(x)$ at all these nodes are denoted as $[f(x_0), f(x_1), \ldots, f(x_n)]$, the function value at any point x within the interval can be interpolated using

$$f(x) = \sum_{i=0}^{n} f(x_i) l_i(x) \qquad (5.7)$$

where l_i is known as the Lagrange interpolator corresponding to node i. It is expressed as follows:

$$l_i(x) = \prod_{i=0,j=1,j\neq i}^{n} \frac{x - x_j}{x_i - x_j}$$

$$= \frac{(x - x_1)(x - x_2)\cdots(x - x_{j-1})(x - x_j)(x - x_{j+1})\cdots(x - x_n)}{(x_i - x_1)(x_i - x_2)\cdots(x_i - x_{j-1})(\quad 1 \quad)(x_i - x_{j+1})\cdots(x_i - x_n)}$$

(5.8)

A Lagrange interpolator, $l_i(x)$, is also a type of the **shape functions** used in the FEM often denoted as $N_i(x)$.

5.4.2 *Properties of Lagrange interpolation*

With this unique construction in Eq. (5.8), the basis function of a Lagrange interpolator has the following features:

1. The interpolator $l_i(x)$ satisfies

$$l_i(x_j) = \delta_{ij} = \begin{cases} 1 & \text{when } i = j \\ 0 & \text{when } i \neq j \end{cases}$$

(5.9)

This is also known as the **Delta function property**. This property can be stated as follows: *the interpolator has value 1 at its home node and zero at all the remote nodes.*

Due to this, Eq. (5.7) can produce the function values at all these nodes in the interval $[a, b]$. This is known as node-passing interpolation [1] and can be proven as follows:

$$f(x_j) = \sum_{i=0}^{n} f(x_i) l_i(x_j) = \sum_{i=0}^{n} f(x_i) \delta_{ij} = f(x_j)$$

(5.10)

2. The sum of all $l_i(x)$ at any point in the interval gives unity:

$$\sum_{i=0}^{n} l_i(x) = 1$$

(5.11)

This is known as the **partition of unity property** [3]: each of the interpolators is a part of the unit. Due to this, Eq. (5.7) can produce the function exactly if the function is a constant $f(x) = c$. This can be shown as follows:

$$f(x) = \sum_{i=0}^{n} \underbrace{f(x_i)}_{c} l_i(x) = c \underbrace{\sum_{i=0}^{n} l_i(x)}_{1} = c \tag{5.12}$$

3. The highest degree (also called order) of a polynomial is n. If the function $f(x)$ is a polynomial with a degree lower than or equal to n, the interpolation will produce exactly its values at any point in the interval. This is known as the reproducibility property. A proof can be found in Ref. [3]. We will "prove" this using examples later.
4. If the function $f(x)$ is a polynomial with a degree higher than n, or any other form of function, the interpolation will be an approximation.
5. Note that Eq. (5.7) may also be used for extrapolation to approximate the function values beyond $[a, b]$. However, extra care is needed when doing so because such an approximation can be quite far off.

5.4.3 *Proof Lagrange interpolators as basis functions*

The proof that Lagrange interpolators can be used as the basis functions for polynomials of order n is quite straightforward. We first need to prove that these interpolators are linearly independent.

We note each $l_i(x)$ with order of n is a polynomial with $n+1$ coefficients. From its property given in Eq. (5.9), each $l_i(x)$ with order of n must satisfy $n + 1$ different (assuming all these nodes are distinct) conditions at these $n + 1$ nodes. The number of conditions is exactly the same as the number of these coefficients, the Vandermonde matrix is not singular, and the right-hand side is non-zero. Therefore, each $l_i(x)$ is uniquely determined by its own set of $n + 1$ conditions. Hence, it is linearly independent of any other $l_j(x)$ ($j \neq i$). Therefore, $l_i(x)$ ($i = 0, 1, \ldots, n$) are qualified as a set of basis functions for a polynomial space of order n.

It is clear that the linear independence is achieved by letting each basis function having its own home node. In this case, the order of the basis functions can be the same, as will be shown in the following examples.

5.4.4 *Python codes for generating Lagrange interpolators*

The following is a Sympy code for Lagrange interpolations:

```python
def LagrangeP(x, xi):

    '''Create shape function matrix N(x)=[l_0(x), l_1(x),...l_degree(x)
        in a list for a 1D interval (element) [a=x_0, x_1, ...,  b=x_n],
        using the Lagrange interpolators.
    '''
    N = []                                    # L_0(x) is a shape function N(x)
    for i in range(len(xi)):
        nodes = [n for n in range(len(xi))]
        del nodes[i]
        N.append(np.prod([(x-xi[k])/(xi[i]-xi[k]) for k in nodes]))

    return N
```

First, let us use only two nodes in the domain and create two Lagrange interpolators for each of these two nodes:

```python
x = sp.symbols('x')                    # define a symbolic variable

nodes = [0, 1]        # x coordinate values at these two nodes
Nx = sp.Matrix([simplify(s) for s in LagrangeP(x, nodes)]).T
Nx
```

$$[1 - x \quad x]$$

We obtained a list of two interpolator functions, all of degree 1. They all satisfy the properties mentioned earlier. This is the so-called shape function for domain $[0, 1]$. Assume that a function has values at these two nodes of 1.0 and 2.0, we can use Eq. (5.7) to obtain the expression of the function:

```python
fi = sp.Matrix([1, 2])
Nx@fi
```

$$[x + 1]$$

We obtained a function that is a linear function within the domain $[0, 1]$, as expected.

Let us now create Lagrange shape functions of degree 3, use P3_4_points obtained in the previous section to generate the function, and then compare it with the polynomial P3(x).

```
1  nodes =[1, 2, 3, 4]      # coordinate values at these 4 nodes
2  Nx = sp.Matrix([simplify(s) for s in LagrangeP(x, nodes)]).T
3  Nx
```

$$\left[-\frac{(x-4)(x-3)(x-2)}{6} \quad \frac{(x-4)(x-3)(x-1)}{2} \quad -\frac{(x-4)(x-2)(x-1)}{2} \quad \frac{(x-3)(x-2)(x-1)}{6}\right]$$

We obtained four interpolators (shape functions). Using Eq. (5.7), we have the following:

```
1  L3 = Nx@sp.Matrix(P3_4_points)
2  L3 = L3.expand().evalf(3)
3  L3
```

$$\left[-24.0x^3 + 26.0x^2 - 9.0x + 8.0\right]$$

This is exactly the function P3 we obtained in the previous section using monomials. This demonstrates that the Lagrange basis functions of degree n can uniquely produce polynomial functions of the same degree. This is because they all live in the same polynomial space of order n. If monomial basis functions are used to express a function rather than polynomials, it would be an approximation.

One can create Lagrange basis functions of any desired degree. Let us create interpolator functions of degree 8:

```
1  degree = 8                              # the degree of L_i(x)
2  nodes = [n for n in range(degree+1)]
3  print(f'nodes: {nodes}')
4  l_1to9 = LagrangeP(x, nodes)            # print only the 1st L_0(x)
5  l_1to9[0].simplify()
```

nodes: [0, 1, 2, 3, 4, 5, 6, 7, 8]

$$\frac{(x-8)(x-7)(x-6)(x-5)(x-4)(x-3)(x-2)(x-1)}{40320}$$

There are a total of nine interpolator functions. The one printed above is for node 0. It can be seen clearly when $x = 0$, it is 1, and when $x = 1, 2, \ldots, 8$, it is 0. It satisfies the Delta function property.

5.4.5 *Lagrange basis functions in natural coordinate system*

One can also use the natural coordinates, which is more convenient to use in many applications, say in FEM. Let us create some such shape function using the following code:

```
1  ξ=symbols('ξ')                           # define natrual coordinate
2
3  # Linear: using two nodes:
4  nodes = [-1,1]                   # natrual coordinates values at these nodes
5  N_1st = [sp.factor(s) for s in LagrangeP(ξ, nodes)]
6  print(f'l_0 at x_0 = {N_1st[0].subs(ξ,nodes[0])}')    # delta property
7  print(f'l_0 at x_1 = {N_1st[0].subs(ξ,nodes[1])}')
8  print(f'Sum of all l_s = {sum(N_1st).simplify().evalf(4)}')    # unity
9  N_1st
```

l_0 at x_0 = 1

l_0 at x_1 = 0

Sum of all l_s = 1.000

$$\left[-\frac{\xi-1}{2}, \ \frac{\xi+1}{2} \right]$$

```
1  # Quadratic: using three nodes in a symmetric domain:
2  nodes = [-1, 0, 1]           # natrual coordinate values at these nodes
3  N_2nd = [sp.factor(s) for s in LagrangeP(ξ, nodes)]
4  print(f'l_1 at x_0 = {N_2nd[1].subs(ξ,nodes[0])}')    # delta property
5  print(f'l_1 at x_1 = {N_2nd[1].subs(ξ,nodes[1])}')
6  print(f'Sum of all l_s = {sum(N_2nd).simplify().evalf(4)}')    # unity
7  N_2nd
```

l_1 at x_0 = 0

l_1 at x_1 = 1

Sum of all l_s = 1.000

$$\left[\frac{\xi(\xi-1)}{2}, \ -(\xi-1)(\xi+1), \ \frac{\xi(\xi+1)}{2} \right]$$

```
1  # Quadratic: using three nodes in domain [1, 2, 3]:
2  nodes = [1,2,3]              # natrual coordinate values at these nodes
3  N_2nd = [sp.factor(s) for s in LagrangeP(ξ, nodes)]
4  print(f'l_1 at x_0 = {N_2nd[1].subs(ξ,nodes[0])}')    # delta property
5  print(f'l_1 at x_1 = {N_2nd[1].subs(ξ,nodes[1])}')
6  print(f'Sum of all l_s = {sum(N_2nd).simplify().evalf(4)}')    # unity
7  N_2nd
```

l_1 at x_0 = 0

l_1 at x_1 = 1

Sum of all l_s = 1.000

$$\left[\frac{(\xi-3)(\xi-2)}{2}, \ -(\xi-3)(\xi-1), \ \frac{(\xi-2)(\xi-1)}{2} \right]$$

5.4.6 *Python codes for the distribution of Lagrange interpolators*

Let us write the following codes to compute and plot the distribution of the Lagrange interpolators used for first-, second-, third-, and fourth-order interpolations:

```python
1  X = np.linspace(-1., 1., 200)              # rcnge in real axis for plot
2
3  Labels=['1st order','2nd order','3rt order','4th order']
4  LNs=['$l_0$','$l_1$','$l_2$','$l_3$', '$l_4$']
5
6  Nodes = [[-1, 1], [-1, 0, 1], [-1,-1/3.,1/3.,1], [-1, -.5, 0, .5, 1]]
7
8  plt.rcParams.update({'font.size': 6})
9  plt.rcParams["figure.autolayout"] = True
10 fig_s = plt.figure(figsize=(6,4))
11
12 for i in range(len(Nodes)):
13     ax = fig_s.add_subplot(2,2,i+1)
14     fs = [lx for lx in LagrangeP(x, Nodes[i])]
15     lambda_f = [lambdify(x,f) for f in fs]
16     for n in range(len(fs)):
17         ax.plot(X, lambda_f[n](X), lw=1, label=LNs[n])
18
19     ax.scatter(Nodes[i], [0]*len(Nodes[i]), c='k', s=7)
20     ax.set_xlabel('x')
21     if i==0 or i==2: ax.set_ylabel('$l(x)$')
22     ax.legend() #loc='center right', bbox_to_anchor=(1, 0.5))
23     plt.title(Labels[i])
24
25 fig_s.tight_layout()
26 plt.savefig('images/shapeFunctions1_4.png',dpi=500,bbox_inches='tight')
27 #plt.show()
```

Figure 5.1 shows the following:

1. All these interpolators satisfy the Delta function property: the interpolator has value 1 at its home node and zero at all the remote nodes.
2. All these interpolators satisfy the partitions of unity property: the sum of all the interpolators at any x inside the domain gives 1.
3. They are all smooth inside the domain, and higher degree ones vary more inside the domain and can have steep slopes near the boundary nodes.

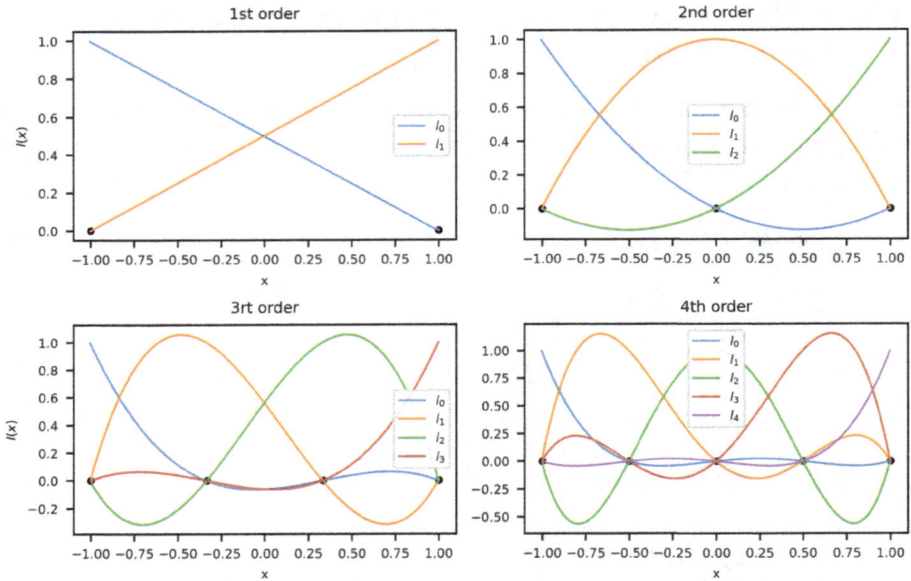

Figure 5.1. Lagrange interpolators obtained using different numbers of nodes. These interpolators are used as basis functions (or shape functions) with different orders.

5.5 Chebyshev polynomials

5.5.1 *Formulation and recursive formula*

Chebyshev polynomials are linearly independent and, hence, are also qualified as the basis functions for approximating a function. This family of functions is well studied since it was formulated by Chebyshev in the late 1800s. It is one of the orthogonal function groups in the integral sense, and there are a number of ways to define them, as well summarized in the Wikipedia page on Chebyshev_polynomials. Here, we use the recurrence formulation that gives a straightforward definition:

$$T_0(x) = 1 \qquad\qquad \text{for } n = 0$$
$$T_1(x) = x \qquad\qquad \text{for } n = 1 \qquad (5.13)$$
$$T_n(x) = 2x\,T_{n-1}(x) - T_{n-2}(x) \quad \text{for } n > 1$$

This gives the whole family of the first kind of Chebyshev polynomials. It is denoted as T because the name Chebyshev is Tchebyshev in French.

Using Eq. (5.13), we can define a Numpy function to generate as many terms as we want.

5.5.2 *Some essential properties of Chebyshev polynomials of the first kind*

The construction in Eq. (5.13) ensures the following:

1. $T_n(1) = 1$. This is clearly true for the first two equations in Eq. (5.13). When setting $x = 1$ in the third equation in Eq. (5.13), we get

$$T_n(1) = 2\,T_{n-1}(1) - T_{n-2}(1); \quad \text{for } n > 1 \qquad (5.14)$$

   This simple recursive operation gives $T_n(1) = 1$ for all n.

2. $T_n(x)$ is always one order higher than $T_{n-1}(x)$. Thus, all these $T_n(x)$ are linearly independent of each other, implying that any one of them cannot be expressed as a linear combination of the rest (Note that the third equation in Eq. (5.13) is not a linear combination). The linear independence among these basis functions is achieved by having different orders similar to the monomial basis functions. Hence, they are qualified as the basis functions for approximating any continuous function.

3. The polynomials are complete, implying that bases $T_0(x), T_1(x),$ $T_2(x), \ldots, T_n(x)$ are equivalent to the monomial basis functions studied in the previous section: $1, x, x^2, \ldots, x^n$. They both span the same space of polynomials of degree n (and of course live in the same space). This gives the ability to reproduce any polynomial with order of n.

5.5.3 *Chebyshev polynomials of the second kind*

The second kind of Chebyshev polynomials have the following the recurrence formula:

$$\begin{aligned}
U_0(x) &= 1 & &\text{for } n = 0 \\
U_1(x) &= 2x & &\text{for } n = 1 \qquad (5.15) \\
U_n(x) &= 2x\,U_{n-1}(x) - U_{n-2}(x) & &\text{for } n > 1
\end{aligned}$$

The following Python function can generate them:

5.5.4 *Python codes to generate Chebyshev polynomials*

```python
1  def cheby_T(n, x):
2
3      '''Generate the first kind of Chebyshev polynomials of degree (n-1)
4          in Sympy expression.
5      Inputs:
6          n: integer, (n-1) is the order of the polynomial.
7          x: symbolic variable
8      Return:
9          the nth term of the Chebyshev polynomial
10      '''
11      if   n == 0: return sp.S.One
12      elif n == 1: return x
13      else:        return 2*x*cheby_T(n-1,x)-cheby_T(n-2,x)
```

The following Sympy code generates the first n terms:

```python
1  x = symbols('x')        # define a symbolic variable
2  n = 5                   # generate the first n terms
3  for i in range(n):
4      print(f"T_{i}(x) = {cheby_T(i, x).expand()}")
```

```
T_0(x) = 1
T_1(x) = x
T_2(x) = 2*x**2 - 1
T_3(x) = 4*x**3 - 3*x
T_4(x) = 8*x**4 - 8*x**2 + 1
```

The following equations list the first 11 terms for easy examination:

$$T_0(x) = 1$$

$$T_1(x) = x$$

$$T_2(x) = 2x^2 - 1$$

$$T_3(x) = 4x^3 - 3x$$

$$T_4(x) = 8x^4 - 8x^2 + 1$$

$$T_5(x) = 16x^5 - 20x^3 + 5x$$

$$T_6(x) = 32x^6 - 48x^4 + 18x^2 - 1$$

$$T_7(x) = 64x^7 - 112x^5 + 56x^3 - 7x$$

$$T_8(x) = 128x^8 - 256x^6 + 160x^4 - 32x^2 + 1$$

$$T_9(x) = 256x^9 - 576x^7 + 432x^5 - 120x^3 + 9x$$

$$T_{10}(x) = 512x^{10} - 1280x^8 + 1120x^6 - 400x^4 + 50x^2 - 1 \qquad (5.16)$$

One may use the built-in Sympy function Chebyshev $t(i, x)$ to generate the same. One may observe the pattern of the coefficients, especially the leading-order terms and the constant terms. These coefficients belong to OEIS: A028297.

The following code generates the Chebyshev polynomial of the second kind:

```python
def cheby_U(n, x):

    '''Generate the second kind Chebyshev polynomials of degree (n-1)
        in Sympy expression.
    Inputs:
        n: integer, (n-1) is the order of the polynomial.
        x: symbolic variable
    Return:
        the nth term of the Chebyshev polynonial
    '''
    if   n == 0: return S.One
    elif n == 1: return 2*x
    else:        return 2*x*cheby_U(n-1,x)-cheby_U(n-2,x)
```

The following Sympy code generates the first n terms:

```python
n = 5                    # generate the first n terms
for i in range(n):
    print(f"U_{i}(x) = {cheby_U(i, x).expand()}")
```

```
U_0(x) = 1
U_1(x) = 2*x
U_2(x) = 4*x**2 - 1
U_3(x) = 8*x**3 - 4*x
U_4(x) = 16*x**4 - 12*x**2 + 1
```

The following equations list the first 11 terms:

$$U_0(x) = 1$$

$$U_1(x) = 2x$$

$$U_2(x) = 4x^2 - 1$$

$$U_3(x) = 8x^3 - 4x$$

$$U_4(x) = 16x^4 - 12x^2 + 1$$

$$U_5(x) = 32x^5 - 32x^3 + 6x$$

$$U_6(x) = 64x^6 - 80x^4 + 24x^2 - 1$$

$$U_7(x) = 128x^7 - 192x^5 + 80x^3 - 8x$$

$$U_8(x) = 256x^8 - 448x^6 + 240x^4 - 40x^2 + 1$$

$$U_9(x) = 512x^9 - 1024x^7 + 672x^5 - 160x^3 + 10x$$

$$U_{10}(x) = 1024x^{10} - 2304x^8 + 1792x^6 - 560x^4 + 60x^2 - 1 \qquad (5.17)$$

One may use the built-in Sympy function Chebyshev $u(i, x)$ to generate the same. Note the pattern of the coefficients, especially the leading-order terms and the constant terms. These coefficients belong to OEIS: A053117.

These two kinds of Chebyshev polynomials are not independent and have the following relationship:

$$T_{n+1}(x) = x \, T_n(x) - (1 - x^2) \, U_{n-1}(x)$$

$$U_{n+1}(x) = x \, U_n(x) + T_{n+1}(x) \qquad (5.18)$$

$$T_n(x) = \frac{1}{2}(U_n(x) - U_{n-2}(x))$$

Therefore, we can generate one kind from another. The first kind is more often used, but sometimes, it is more convenient to use the second kind, for example, in numerical integrations and in solving some differential equations with some boundary conditions. In Sympy, both of these are equally easy to generate.

5.5.5 *Roots of Chebyshev polynomials*

A Chebyshev polynomial of either kind with degree n has n distinct roots, called Chebyshev roots, in the interval $[-1, 1]$. These roots are sometimes called Chebyshev nodes and have applications in function approximation, solving differential equations, etc. Let us compute the roots of Chebyshev polynomials of both kinds using the following code:

```
1  #help(sp.chebyshevt_root)          # one may use this to get the user guide
```

```
1  N = 5                               # number of Chebyshev polynomials
2  t_rs = sp.chebyshevt_root; u_rs = sp.chebyshevu_root
3
4  T_rs=[[t_rs(n,k).evalf() for k in range(n)] for n in range(1,N+1)]
5  U_rs=[[u_rs(n,k).evalf() for k in range(n)] for n in range(1,N)]
6
7  # flatten the lists in a list & remove the duplicated roots.
8  T_rs = list(set(list(itertools.chain.from_iterable(T_rs))))
9  U_rs = list(set(list(itertools.chain.from_iterable(U_rs))))
10
11 gr.printl(np.sort(T_rs), 'roots of Tn:\n')
12 gr.printl(np.sort(U_rs), 'roots of Un:\n')
```

roots of Tn:
 [-0.9511, -0.9239, -0.866, -0.7071, -0.5878, -0.3827, 0.0, 0.3827, 0.5878,
0.7071, 0.866, 0.9239, 0.9511]
roots of Un:
 [-0.809, -0.7071, -0.5, -0.309, 0.0, 0.309, 0.5, 0.7071, 0.809]

These roots for T_n can also be found using the following formula:

$$\text{root}_k = \cos \frac{(2k+1)\pi}{2n+2} \quad \text{for} \quad 0 \leq k \leq n \tag{5.19}$$

The following is a Python code using this formula:

```
1  N = 5                                    # number of Chebyshev polynomials
2  T_rs=[[cos((2*k+1)*np.pi/(2*n+2)) for k in range(n+1)]
3                               for n in range(1,N)]
4
5  # flatten the lists in a list & remove the duplicated roots.
6  T_rs = list(set(list(itertools.chain.from_iterable(T_rs))))
7  gr.printl(np.sort(T_rs), 'roots of Tn:\n')
```

roots of Tn:
 [-0.9511, -0.9239, -0.866, -0.7071, -0.5878, -0.3827, 0.0, 0.3827, 0.5878,
0.7071, 0.866, 0.9239, 0.9511]

5.5.6 *Distribution of Chebyshev polynomials*

We write the following code function to compute and plot the distribution of polynomial basis functions. This code will be used for Chebyshev polynomials and also for other basis functions to be discussed later.

```
1  def plot_PolyBasis(p_name='tmp'):
2
3      '''Compute, save and plot the distribution of four basis functions
4      defined in Sympy function object Px, together with the roots.
5      '''
6      plt.rcParams.update({'font.size': 6})
7      fig_s = plt.figure(figsize=(5,3))
8
9      for i in range(len(ylables)):
10
11         ax = fig_s.add_subplot(1,1,i+1)
12         for n in range(N):
13             f = lambdify(x,Px(n,x))
14             ax.plot(X, [f(xi) for xi in X], lw=1, label='n='+str(n))
15
```

```
16          ax.scatter(roots, [0]*len(roots), c='r',s=7, label='roots')
17          ax.set_xlabel('x')
18          ax.set_ylabel(ylables[i]); ax.set_ylim(y_min, y_max)
19
20          ax.grid(c='r', linestyle=':', linewidth=0.5)
21          ax.axvline(x=0, c="k", lw=0.6); ax.axhline(y=0, c="k", lw=0.6)
22          ax.legend()        #Loc='center right', bbox_to_anchor=(1, 0.5))
23
24      plt.savefig('images/'+p_name+'Basis.png',dpi=500,bbox_inches='tight
25      plt.show()
```

The following code sets and plots four Chebyshev basis functions of the first kind:

```
1  N = 6                              # number of Chebyshev polynomials
2  a, b = -1, 1                       # domain for the basis functions
3  y_min, y_max = -1.1, 1.1           # control y range of the plot
4  X = np.linspace(a, b, 200)         # x range in real axis
5
6  ylables = ['Chebyshev polynomials, $T_n(x)$']
7
8  P_type = 'Chebyshev'
9  Px = sp.chebyshevt                 # Class object
10 roots = T_rs                       # Use the roots computed earlier
11
12 plot_PolyBasis(p_name=P_type)
```

Figure 5.2 shows the following:

- Chebyshev basis functions of the first kind are all bounded by ± 1 in the domain of $[-1, 1]$.

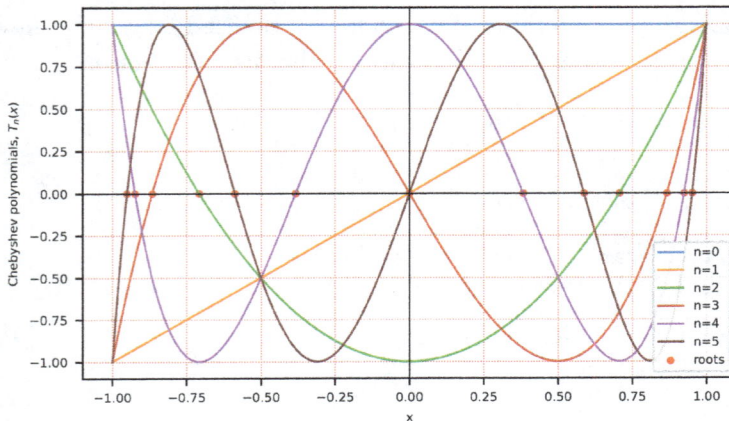

Figure 5.2.   Chebyshev polynomials of first kind with different orders.

- They all have the same value of 1 at $x = 1$.
- Functions with odd degrees are anti-symmetric. When used for approximating odd functions, only these odd basis functions are needed. Functions with even degrees are symmetric. When approximating even functions, only these even bases are needed. When the function to be approximated is asymmetric, we shall use both.
- The roots of any Chebyshev basis function are distinct. The roots of all Chebyshev basis functions have multiplicity at $x = 0$.

The following code sets and plots four Chebyshev basis functions of the second kind:

```
1  y_min, y_max = -4.1, 4.1
2  ylables = ['Chebyshev polynomials, $U_n(x)$']
3
4  P_type = 'Chebyshevu'
5  Px = sp.chebyshevu                          # Class object
6  roots = U_rs
7  plot_PolyBasis(p_name=P_type)
```

Figure 5.3 shows the following for the second kind of Chebyshev polynomials:

- For high degrees, Chebyshev basis functions of the second kind vary fast near the boundary of the domain of $[-1, 1]$.
- Functions with odd degrees are anti-symmetric. Functions with even degrees are symmetric.

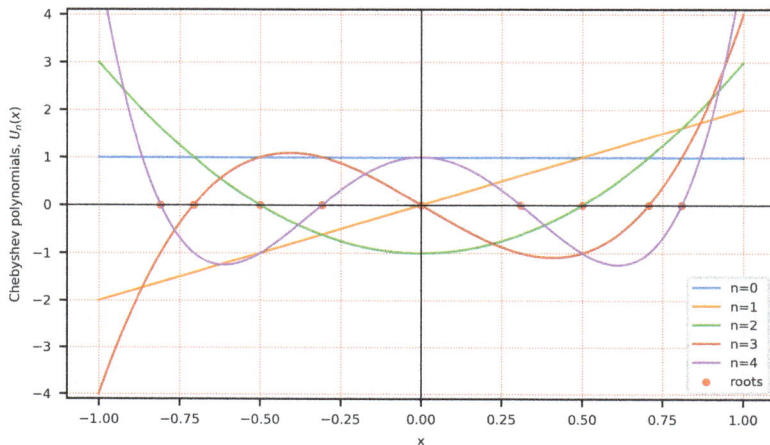

Figure 5.3. Chebyshev polynomials of second kind with different orders.

- The roots of any Chebyshev basis function are distinct. The roots of all Chebyshev basis functions can have multiplicity in the domain.

5.5.7 *Conversion between Chebyshev and primitive polynomials*

Chebyshev polynomials can be converted to primitive polynomials and vice versa. Python provides convenient methods to do so. The conversion is unique because they all live in the same polynomial space of the same order.

```
1  import numpy.polynomial as Poly
2  cheb = Poly.Chebyshev([5,2,3])        # create a Chebyshev poly
3  cheb
```

$$x \mapsto 5.0\,T_0(x) + 2.0\,T_1(x) + 3.0\,T_2(x)$$

```
1  cheb.convert(kind=Poly.Polynomial)    # us convert() method
```

$$x \mapsto 2.0 + 2.0\,x + 6.0\,x^2$$

```
1  P_coef = Poly.chebyshev.cheb2poly([5,2,3])    # use cheb2poly()
2  P_coef
```

```
array([2., 2., 6.])
```

```
1  Poly.chebyshev.poly2cheb(P_coef)      # convert back
```

```
array([5., 2., 3.])
```

5.6 Legendre polynomials

5.6.1 *Definition, Bonnet's recursive formula*

Legendre polynomials are the other classical orthogonal polynomials (in the integral sense). They are named after Adrien-Marie Legendre (1782) and are widely used in computational methods. They are well studied, and there are a number of ways to define them. Here, we use Bonnet's recursive formula:

$$P_0(x) = 1 \quad \text{for } n = 0$$

$$P_1(x) = x \quad \text{for } n = 1 \tag{5.20}$$

$$P_n(x) = \frac{1}{n}\big((2n-1)xP_{n-1}(x) - (n-1)P_{n-2}(x)\big) \quad \text{for } n > 1$$

where n is a positive integer.

5.6.2 *Some essential properties*

This construction ensures the following:

1. $P_n(1) = 1$. This is obvious in the first two equations in Eq. (5.20). Setting $x = 1$, the third equation in Eq. (5.20) becomes

$$P_n(1) = \frac{1}{n}((2n-1)P_{n-1}(1) - (n-1)P_{n-2}(1)) \quad \text{for } n > 1 \qquad (5.21)$$

This simple recursive operation gives $P_n(1) = 1$ for all n.
2. All these $P_n(x)$ are linearly independent of each other. $P_n(x)$ is always one order higher than $P_{n-1}(x)$. The linear independence among these basis functions is achieved by having different orders similar to the Chebyshev basis functions. Thus, they are qualified as the basis functions for approximating any continuous function.
3. The polynomials are complete, implying that basis functions $P_0(x)$, $P_1(x), P_2(x), \ldots, P_n(x)$ are equivalent to the monomial basis functions studied in the previous chapter: $1, x, x^2, \ldots, x^n$. They span the same space. This gives the ability to reproduce any polynomial with order of n.

We examine all these using the codes given in the following section. First, let us write a code to generate these functions.

5.6.3 *Python codes to generate Legendre polynomials*

```
 1  def LegendreR(n, x):
 2
 3      '''Generate the Legendre polynomials of degree (n-1) using
 4          Bonnet's recursion formula, in Sympy expression.
 5      Inputs:
 6          n: integer, (n-1) is the order of the polynomial.
 7          x: symbolic variable
 8      Return:
 9          the nth term of the Legendre polynomial
10      '''
11      if   n == 0: return sp.S.One
12      elif n == 1: return x
13      else:
14          leg = ((2*n-1)*x*LegendreR(n-1,x)-(n-1)*LegendreR(n-2,x))/n
15          return leg.expand()
```

The following Sympy code generates the first n terms:

```python
1  x = sp.symbols('x')                    # define a symbolic variable
2  n = 6                                  # generate the first n terms
3  for i in range(n):
4      print(f"P_{i}(x) = {LegendreR(i, x)}")
```

```
P_0(x) = 1
P_1(x) = x
P_2(x) = 3*x**2/2 - 1/2
P_3(x) = 5*x**3/2 - 3*x/2
P_4(x) = 35*x**4/8 - 15*x**2/4 + 3/8
P_5(x) = 63*x**5/8 - 35*x**3/4 + 15*x/8
```

One may also use the Sympy built-in functions to generate Legendre polynomials:

```python
1  from sympy.polys.orthopolys import legendre_poly
2  n = 6
3  legendre_Ps = [legendre_poly(i, x) for i in range(n)]    # first n terms
4  legendre_Ps    # help(sp.polys.orthopolys)
```

$$\left[1,\ x,\ \frac{3x^2}{2} - \frac{1}{2},\ \frac{5x^3}{2} - \frac{3x}{2},\ \frac{35x^4}{8} - \frac{15x^2}{4} + \frac{3}{8},\ \frac{63x^5}{8} - \frac{35x^3}{4} + \frac{15x}{8}\right]$$

5.6.4 *Roots of the Legendre polynomials*

The roots of Legendre polynomials are useful in computational methods, including approximation of functions, quadrature rules for functions, and solving eigenvalue problems.

The following code finds the roots of any Legendre polynomials, using the sp.solve(). We use nested loops to find all the roots of all Legendre polynomials for a given degree and then remove the duplicated ones:

```python
1  N = 6                                  # number of Legendre polynomials
2  x = sp.Symbol('x')                     # sympolid variable
3
4  roots = [[r.evalf() for r in sp.solve(sp.legendre(n, x), x)]
5              for n in range(N)]    # Find the roots of all Pn
6
7  # flatten the lists in a list and remove the duplicated roots.
8  roots = list(set(list(itertools.chain.from_iterable(roots))))
9  gr.printl(roots[:9], 'roots of Pn:\n')
```

```
roots of Pn:
  [0.0, 0.5385, -0.9062, 0.9062, -0.5774, -0.8611, 0.34, 0.7746, -0.7746]
```

Note that a P_n has n roots, and these roots are all distinct.

5.6.5 *Distribution of the Legendre polynomials*

We write the following code to plot the distribution of Legendre polynomials together with the roots using our plot_PolyBasis() defined earlier:

```
 1  N = 6                              # number of Chebyshev polynomials
 2  a, b = -1, 1
 3  y_min, y_max = -1.1, 1.1
 4  X = np.linspace(a, b, 200)                      # range in real axis
 5
 6  ylables = ['Legendre polynomials, $P_n(x)$']
 7
 8  P_type = 'Legendre'
 9  Px = sp.legendre                                # Class object
10  plot_PolyBasis(p_name=P_type)
```

We note the following from Fig. 5.4:

- Legendre basis functions of the first kind are all bounded by ± 1 in the domain of $[-1, 1]$.
- They all have the same value of 1 at $x = 1$.
- Functions with odd degrees are anti-symmetric. When used for approximating odd functions, only these odd basis functions are needed. Functions with even degrees are symmetric. When approximating even functions, only these even bases are needed. When the function to be approximated is asymmetric, we shall use both.
- The roots of any Legendre basis function are distinct. The roots of all Legendre basis functions have multiplicity at $x = 0$.

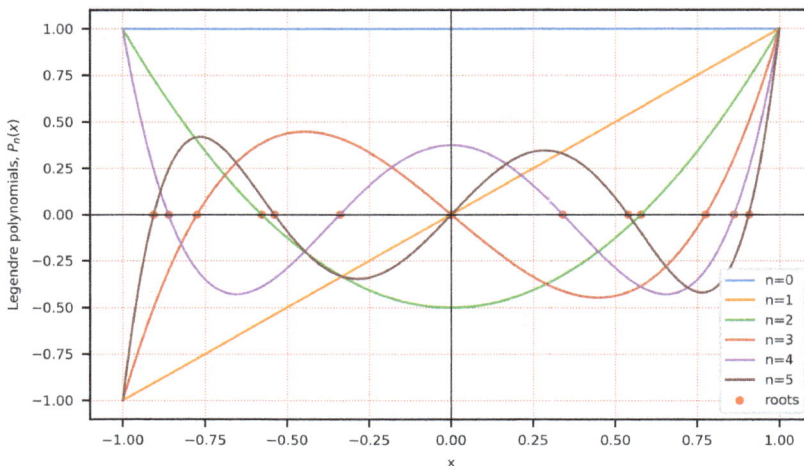

Figure 5.4. Legendre polynomials with different orders and their roots.

These findings are largely similar to those we found for the Chebyshev polynomials.

5.6.6 *Conversion between Legendre and primitive polynomials*

Legendre polynomials can be converted to primitive polynomials and vice versa. Python provides convenient methods to do so:

```python
import numpy.polynomial as Poly
leg = Poly.Legendre([5, 2, 3])        # generate a Legendre using coeffs.
leg
```

$$x \mapsto 5.0\, P_0(x) + 2.0\, P_1(x) + 3.0\, P_2(x)$$

```python
leg.convert(kind=Poly.Polynomial)                # use convert() method
```

$$x \mapsto 3.5 + 2.0\, x + 4.5\, x^2$$

```python
P_coef = Poly.legendre.leg2poly([5,2,3])    # use Leg2poly() method
P_coef
```

```
array([3.5, 2. , 4.5])
```

```python
Poly.legendre.poly2leg(P_coef)              # use poly to Legendre
```

```
array([5., 2., 3.])
```

Note that the Legendre polynomials can also be defined in the domain of $[0, 1]$. Readers can easily do so by slightly modifying the codes given above.

5.7 Laguerre polynomials

5.7.1 *Definition, the recursive formula*

Laguerre polynomials are also classical orthogonal polynomials. They are named after Edmond Laguerre (1834–1886), and are the solutions of Laguerre's differential equation. They are used in computational methods. They are well studied, and there are a number of ways to define them, and usually their domain is $[0, \infty)$ and codomain is $(-\infty, \infty)$. Here, we use the recursive formula:

$$L_0(x) = 1 \qquad \text{for } n = 0$$
$$L_1(x) = -x + 1 \qquad \text{for } n = 1 \qquad (5.22)$$
$$L_n(x) = \frac{1}{n}\big((2n - 1 - x)L_{n-1}(x) - (n - 1)L_{n-2}(x)\big) \quad \text{for } n > 1$$

where n is a positive integer.

5.7.2 *Some essential properties*

This construction ensures that

1. $L_n(0) = 1$ for all n. This is observed for the $n = 0$ and $n = 1$. For $n > 1$, the recursive equation in Eq. (5.22) gives $L_n(0) = 1$ for all n.
2. All these $L_n(x)$ are linearly independent of each other. $L_n(x)$ is always one degree higher than $L_{n-1}(x)$. Thus, they are qualified as the basis functions for approximating any continuous function.
3. The polynomials are complete, implying that basis functions $L_0(x), L_1(x)$, $L_2(x), \ldots, L_n(x)$ are equivalent to the monomial basis functions studied in the previous chapter: $1, x, x^2, \ldots, x^n$. This gives the ability to reproduce any polynomial with order of n.

We examine all these, by writing a code to generate these functions.

```
1  def LaguerreR(n, x):
2
3      '''Generate the Laguerre polynomials of degree (n-1) using
4          the recursion formula, in Sympy expression.
5      Inputs:
6          n: integer, (n-1) is the order of the polynomial.
7          x: symbolic variable
8      Return:
9          the nth term of the Laguerre polynomial
10     '''
11     if   n == 0: return sp.S.One
12     elif n == 1: return -x+1
13     else:
14         Lag = ((2*n-1-x)*LaguerreR(n-1,x)-(n-1)*LaguerreR(n-2,x))/n
15         return Lag.expand()
```

```
1  x = symbols('x')        # define a symbolic variable
2  n = 6                    # generate the first n terms
3  for i in range(n):
4      print(f"L_{i}(x) = {LaguerreR(i, x).expand()}")
```

L_0(x) = 1

L_1(x) = 1 - x

L_2(x) = x**2/2 - 2*x + 1

L_3(x) = -x**3/6 + 3*x**2/2 - 3*x + 1

L_4(x) = x**4/24 - 2*x**3/3 + 3*x**2 - 4*x + 1

L_5(x) = -x**5/120 + 5*x**4/24 - 5*x**3/3 + 5*x**2 - 5*x + 1

5.7.3 *Python codes to generate Laguerre polynomials*

The following Sympy code generates the first n terms.

One may also use the Sympy built-in functions to generate Laguerre polynomials.

```
1  from sympy.polys.orthopolys import laguerre_poly
2
3  n = 5
4  laguerre_Ps = [laguerre_poly(i, x) for i in range(n)] # first n
5  laguerre_Ps
```

$$\left[1, \; 1 - x, \; \frac{x^2}{2} - 2x + 1, \; -\frac{x^3}{6} + \frac{3x^2}{2} - 3x + 1, \; \frac{x^4}{24} - \frac{2x^3}{3} + 3x^2 - 4x + 1 \right]$$

5.7.4 *Roots of the Laguerre polynomials*

The roots of the Laguerre polynomials are useful in a number of applications in computational methods, including function approximation, integration, solving differential equations, calculation of probabilities in some stochastic processes, etc.

The following code finds the roots of any Laguerre polynomial, using the sp.solve(). We use nested loops to find all the roots of all Laguerre polynomials, for a given degree.

```
1  from functools import reduce
2  from operator import concat
3  N = 6                              # number of polynomials
4  x = sp.Symbol('x')                 # sympolic variable
5
6  roots = [[sp.re(r.evalf())  for r in sp.solve(sp.laguerre(n, x), x)]
7                      for n in range(N)]   # Find the roots of all Pn
8
9  # flatten the lists in a list and remove the duplicated roots.
10 roots = list(set(reduce(concat, roots)))
11 roots.sort()
12
13 gr.printl(roots[:9], 'roots of Ln:\n')
```

```
roots of Ln:
 [0.2636, 0.3225, 0.4158, 0.5858, 1.0, 1.4134, 1.7458, 2.2943, 3.4142]
```

Note that a L_n has n roots, and these roots are all distinct.

5.7.5 *Distribution of the Laguerre polynomials*

The following code plots the distribution of Laguerre polynomials, together with their roots.

```
1  N = 6                                # number of Chebyshev polynomials
2  a, b = -1.0, 12.
3  y_min, y_max = -10, 5.
4
5  X = np.linspace(a, b, 200)                        # range in real axis
6
7  ylables = ['Laguerre Polynomials $L_n(x)$']
8
9  P_type = 'Laguerre'
10 Px = sp.laguerre                                    # Class object
11 plot_PolyBasis(p_name=P_type)
```

We note the following from Fig. 5.5:

- It is seen that $L_n(0) = 1$ for all n.
- For $n > 1$, $L_n(x)$ changes drastically when $x < 0$. Thus, its domain is set as $[0, \infty)$. It diverges when x approaches infinity.
- Laguerre basis functions are all asymmetric.
- Laguerre basis functions do not have zero root, and the roots get denser near $x = 0$.

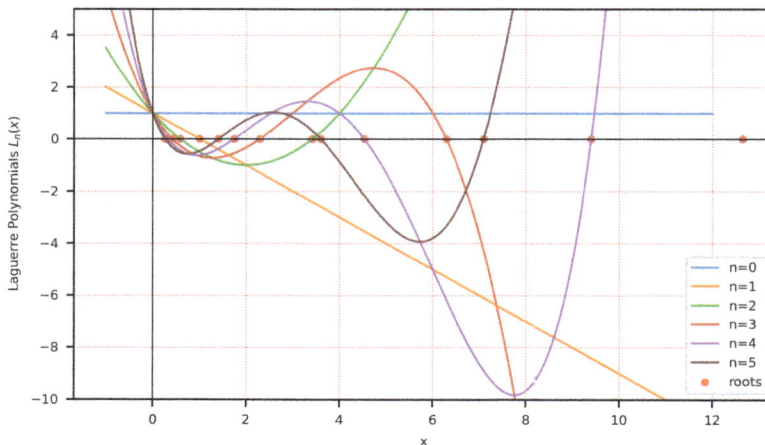

Figure 5.5. Laguerre polynomials with different orders.

These findings are quite different from those we found for the Chebyshev and Legendre polynomials. The domain is different. Also, the weight function used in the orthogonality integral will be different, it needs a fast converging function to control the divergence of these polynomials.

5.7.6 *Conversion between Laguerre and primitive polynomials*

Laguerre polynomials can be converted to primitive polynomials and vice versa. Python provides convenient methods to do so.

```python
import numpy.polynomial as Poly
lag = Poly.Laguerre([5,2,3])        # generate a Legendra using coeffs.
lag
```

$$x \mapsto 5.0\,L_0(x) + 2.0\,L_1(x) + 3.0\,L_2(x)$$

```python
lag.convert(kind=Poly.Polynomial)              # use convert() method
```

$$x \mapsto 10.0 - 8.0\,x + 1.5\,x^2$$

```python
P_coef = Poly.laguerre.lag2poly([5,2,3])       # use Leg2poly() method
P_coef
```

```
array([10. , -8. ,  1.5])
```

```python
Poly.laguerre.poly2lag(P_coef)                 # use poly to Legendre
```

```
array([5., 2., 3.])
```

5.8 Hermite polynomials

5.8.1 *Definition, the recursive formula*

Hermite polynomials are another type of classical orthogonal polynomials. They were originally defined by Pierre-Simon Laplace in 1810 and studied by Pafnuty Chebyshev in 1859 and Charles Hermite in 1864. The Hermite polynomials have a number of applications, including approximation of functions, numerical Gaussian quadrature, and probability analysis.

There are a number of ways to define Hermite polynomials, and the wiki page provides a good summary on this. The domain and codomain of Hermite polynomials are $(-\infty, \infty)$. Here, we use the recurrence formula:

$$
\begin{aligned}
H_0(x) &= 1 && \text{for } n = 0 \\
H_1(x) &= 2x && \text{for } n = 1 \\
H_{n+1}(x) &= 2xH_n(x) - 2nH_{n-1}(x) && \text{for } n > 1
\end{aligned}
\tag{5.23}
$$

where n is a positive integer.

5.8.2 *Some essential properties of $H_n(x)$*

This construction ensures the following:

1. $H_n(x)$ is always one degree higher than $H_{n-1}(x)$, and thus they are linearly independent of each other. Therefore, they are qualified as the basis functions for approximating any continuous function.
2. The polynomials are complete in order, implying that basis functions $H_0(x), H_1(x), H_2(x), \ldots, H_n(x)$ are equivalent to the monomial basis functions studied in the previous chapter: $1, x, x^2, \ldots, x^n$. This gives the ability to reproduce any polynomial with order of n.

5.8.3 *Probabilist's Hermite polynomials*

There is another form of Hermite polynomials known as probabilist's Hermite polynomials defined as

$$
\begin{aligned}
H_{e0}(x) &= 1 && \text{for } n = 0 \\
H_{e1}(x) &= x && \text{for } n = 1 \\
H_{e(n+1)}(x) &= xH_{e(n)}(x) - nH_{e(n-1)}(x) && \text{for } n > 1
\end{aligned}
\tag{5.24}
$$

Both $H_{e(n)}$ and $H_n(x)$ can be generated using the following codes.

5.8.4 *Python codes to generate Hermite polynomials*

```python
def HermiteR(n, x):

    '''Generate the Hermite polynomials of degree (n-1) using
       the recursion formula, in Sympy expression.
    Inputs:
       n: integer, (n-1) is the order of the polynomial.
       x: symbolic variable
    Return:
       the nth term of the Hermite polynomial
    '''
    if   n == 0: return S.One
    elif n == 1: return 2*x
    else:
        Her = 2*x*HermiteR(n-1,x)-2*(n-1)*HermiteR(n-2,x)
    return Her
```

```python
x = symbols('x')        # define a symbolic variable
n = 6                   # generate the first n terms
for i in range(n):
    print(f"H_{i}(x) = {HermiteR(i, x).expand()}")
```

```
H_0(x) = 1
H_1(x) = 2*x
H_2(x) = 4*x**2 - 2
H_3(x) = 8*x**3 - 12*x
H_4(x) = 16*x**4 - 48*x**2 + 12
H_5(x) = 32*x**5 - 160*x**3 + 120*x
```

```python
def HermiteE_R(n, x):

    '''Generate the probabilist's Hermite polynomials of degree (n-1)
       the recursion formula, in Sympy expression.
    Inputs:
       n: integer, (n-1) is the order of the polynomial.
       x: symbolic variable
    Return:
       the nth term of the polynomial
    '''
    if   n == 0: return S.One
    elif n == 1: return x
    else:  return x*HermiteE_R(n-1,x)-(n-1)*HermiteE_R(n-2,x)
```

```
1  x = symbols('x')          # define a symbolic variable
2  n = 6                      # generate the first n terms
3
4  for i in range(n):
5      print(f"H_e{i}(x) = {HermiteE_R(i, x).expand()}")
```

```
H_e0(x) = 1
H_e1(x) = x
H_e2(x) = x**2 - 1
H_e3(x) = x**3 - 3*x
H_e4(x) = x**4 - 6*x**2 + 3
H_e5(x) = x**5 - 10*x**3 + 15*x
```

Our study will focus on $H_n(x)$, which can also be generated using the Sympy built-in functions:

```
1  from sympy.polys.orthopolys import hermite_poly
2  n = 6
3
4  hermite_Ps = [hermite_poly(i, x) for i in range(n)] # first n
5  hermite_Ps
```

$$\left[1,\; 2x,\; 4x^2 - 2,\; 8x^3 - 12x,\; 16x^4 - 48x^2 + 12,\; 32x^5 - 160x^3 + 120x\right]$$

5.8.5 *Roots of the Hermite polynomials*

The following code finds the roots of any Hermite polynomials using the sp.solve(). We use nested loops to find all the roots of all Hermite polynomials for a given degree:

```
1  from functools import reduce
2  from operator import concat
3  N = 6                                  # number of Legendre polynomials
4  x = sp.Symbol('x')                     # sympolid variable
5
6  roots = [[sp.re(r.evalf())  for r in sp.solve(sp.hermite(n, x), x)]
7                      for n in range(N)]       # Find the roots of all Pn
8
9  # flatten the lists in a list & remove the duplicated roots.
10 roots = list(set(reduce(concat, roots)))
11
12 roots.sort()
13
14 gr.printl(roots[:9], 'roots of Pn:\n')
```

```
roots of Pn:
 [-2.0202, -1.6507, -1.2247, -0.9586, -0.7071, -0.5246, 0.0, 0.5246, 0.7071]
```

Note that a P_n has n roots, and these roots are all distinct.

5.8.6 *Distribution of the Hermite polynomials*

We write the following code to plot the distribution of Hermite polynomials together with the roots:

```
1  #help(sp.hermite)
```

```
 1  N = 6                              # number of Chebyshev polynomials
 2  a, b = -2.1, 2.1
 3  y_min, y_max = -42, 42.
 4
 5  X = np.linspace(a, b, 200)         # range in real axis
 6
 7  ylables = ['Hermite Polynomials $L_n(x)$']
 8
 9  P_type = 'Hermite'
10  Px = sp.hermite                    # Class object
11  plot_PolyBasis(p_name=P_type)
```

We note the following from Fig. 5.6:

- The domain and codomain of Hermite basis functions are $(-\infty, \infty)$.
- Since its domain is symmetric, the basis functions made with odd degrees are anti-symmetric. When used for approximating odd functions, only

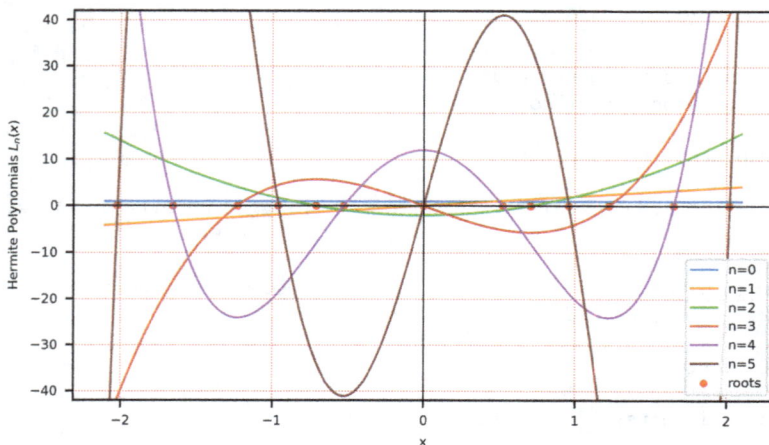

Figure 5.6. Hermite polynomials with different orders.

these odd basis functions are needed. Functions with even degrees are symmetric. When approximating even functions, only these even bases are needed. When the function to be approximated is asymmetric, we shall use both. This is a common feature of all basis functions that are defined in a symmetric domain.

- The roots of any Hermite basis function are distinct. The roots of all Hermite basis functions have multiplicity at $x = 0$.

5.8.7 *Conversion between Hermite and primitive polynomials*

Hermite polynomials can be converted to primitive polynomials and vice versa. Python provides convenient methods to do so:

```
1  import numpy.polynomial as Poly
2  herm = Poly.Hermite([5,2,3])            # generation using coeffs.
3  herm
```

$$x \mapsto 5.0\, H_0(x) + 2.0\, H_1(x) + 3.0\, H_2(x)$$

```
1  herm.convert(kind=Poly.Polynomial)          # use convert() method
```

$$x \mapsto -1.0 + 4.0\, x + 12.0\, x^2$$

```
1  P_coef = Poly.hermite.herm2poly([5,2,3])    # use herm2poly() method
2  P_coef
```

```
array([-1.,   4.,  12.])
```

```
1  Poly.hermite.poly2herm(P_coef)              # converts back
```

```
array([5.,  2.,  3.])
```

5.9 Shape function: Node-based basis functions

One may note that the linear independence of the basis functions of any type (monimial, Chebyshev, Legendre, etc.) is achieved by the differences in the order of each of the basis functions. Therefore, if one would like to increase the number of independent basis functions, one would need to increase the order. This can be problematic because the higher-order polynomials do not behave well: they often vary extremely fast.

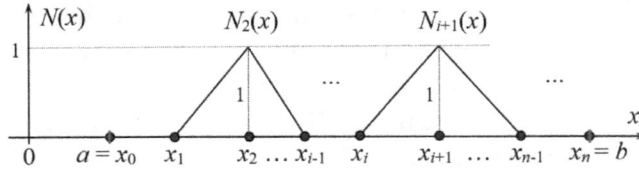

Figure 5.7. Nodal shape functions for a 1D domain $[a, b]$. Each node has a shape function occupying a local region. Two linear shape functions are shown: $N_2(x)$ for node 2 and $N_{i+1}(x)$ for node $i + 1$.

We now introduce another important type of basis functions. They achieve their mutual independence by occupying different regions in the domain.

Consider a 1D domain $[a, b]$. We discretize (also called mesh) it with a set of nodes: $[a = x_0, x_1, \ldots, b = x_n]$. The discretization does not have to be uniform, and the nodes must not overlap. The total number of nodes is $N_n = n + 1$, as shown in Fig. 5.7.

Note that each node has a shape function, and there will be $n + 1$ shape functions. Hence, it is called **nodal shape function**. As each of the shape functions occupies a different local region (also called local support), it is not possible to express any one of the shape functions using the rest, leading to linear independence. This is the major difference of nodal basis from other basis functions discussed in the previous sections. This gives the advantages in solving high-dimensional problems.

5.9.1 *Formulation in the physical coordinate and properties*

Finally, we discuss about a very important function often used in approximation in the finite element method [3], smoothed finite element method [2], and meshfree methods [1]. Our focus here is on the *node-based* shape function for 1D problems.

Consider a 1D bar that is discretized with a set of $n + 1$ nodes. The ith nodal shape function node, $N_i(x)$, can be expressed in the following piecewise form:

$$N_i(x) = \begin{cases} 0 & \text{when } x < x_{i-1} \\ \frac{x - x_{i-1}}{x_i - x_{i-1}} & \text{when } x_{i-1} \leq x < x_i \\ 1 - \frac{x - x_i}{x_{i+1} - x_i} & \text{when } x_i \leq x \leq x_{i+1} \\ 0 & \text{when } x > x_{i+1} \end{cases} \tag{5.25}$$

These N_n shape functions have the following properties of the Lagrange intepolators discussed in Section 4.3.2:

1. **Delta function property:** It is clear that $N_i(x_i) = 1$, $N_i(x_{i-1}) = 0$, $N_i(x_{i+1}) = 0$, varying linearly in $x_{i-1} \leq x \leq x_i$ and $x_i \leq x \leq x_{i+1}$ and zero elsewhere.

2. **Partitions of unity property:** From Eq. (5.25), we have

$$\sum_{i=1}^{N_n} N_i(x) = 1 \tag{5.26}$$

3. **Linear reproducibility:** This means that

$$\sum_{i=1}^{N_n} N_i(x)x_i = x \tag{5.27}$$

5.9.2 *Proof of the linear reproducibility*

The proof of the delta function property and partitions of unity property are trivial. Here, let us prove the linear reproducibility, which requires some work.

First, because of the local support features of these nodal shape functions, we only need to consider a sub-domain between two adjacent nodes (an element), say $[x_{i-1}, x_i]$. In this case, Eq. (5.27) becomes

$$N_{i-1}(x)x_{i-1} + N_i(x)x_i = x \tag{5.28}$$

Next, using Eq. (5.25), the forgoing equation becomes

$$
\begin{aligned}
\left(1 - \frac{x - x_{i-1}}{x_i - x_{i-1}} \right) & x_{i-1} + \left(\frac{x - x_{i-1}}{x_i - x_{i-1}} \right) x_i \\
&= \frac{[(x_i - x_{i-1}) - x + x_{i-1}]x_{i-1}}{x_i - x_{i-1}} + \frac{(x - x_{i-1})x_i}{x_i - x_{i-1}} \\
&= \frac{(x_i - x)x_{i-1}}{x_i - x_{i-1}} + \frac{(x - x_{i-1})x_i}{x_i - x_{i-1}} \\
&= \frac{x_i x_{i-1} - x x_{i-1} + x x_i - x_{i-1}x_i}{x_i - x_{i-1}} \\
&= \frac{-x x_{i-1} + x x_i}{x_i - x_{i-1}} = x
\end{aligned}
\tag{5.29}
$$

5.9.3 *Formulation in the natural coordinate*

Note that the shape function is dimensionless, and hence, we can introduce the so-called **natural coordinate** ξ that is also dimensionless:

$$\xi = \begin{cases} \frac{x-x_i}{x_i-x_{i-1}} & \text{when } x_{i-1} \le x \le x_i \\ \frac{x-x_i}{x_{i+1}-x_i} & \text{when } x_i \le x \le x_{i+1} \end{cases} \tag{5.30}$$

In this case, the domain for non-zero $N_i(\xi)$ becomes $[-1, 1]$, and the formula becomes

$$N_i(\xi) = \begin{cases} 0 & \text{when } \xi < -1 \\ 1 + \xi & \text{when } -1 \le \xi < 0 \\ 1 - \xi & \text{when } 0 \le \xi \le 1 \\ 0 & \text{when } \xi > 1 \end{cases} \tag{5.31}$$

5.9.4 *Python code*

This nodal shape function can be coded in Python as follows:

```python
def node_shape_f(ξ):

    '''Linear nodal shape function in natural coordinate ξ.
    This shape function has 4 pieces.
    '''
    N = np.piecewise(ξ, [ξ<-1, (ξ>=-1)&(ξ<0),  (ξ>=0)&(ξ<=1),  ξ>1],\
                        [0,     lambda ξ: 1+ξ,   lambda ξ: 1-ξ,   0])
    return N
```

The following code plots some nodal shape functions:

```python
plt.rcParams.update({'font.size': 7})
fig, ax = plt.subplots(1,1,figsize=(6,3))
Nn = 11                                          # number of nodes

x_nd = np.array([1+i for i in range(Nn)])
xd = x_nd[1] - x_nd[0]                            # nodal spacing
print(f'Nodes: {x_nd}')

X = np.linspace(-1., 1., 3)              # coordinate with -1 < ξ < 1
fX = node_shape_f(X)                              # N in natrual coordinate

for i in range(1,8):                                   # plot nodal Ns
    ax.plot(X+i*xd, fX, lw=.9)

ax.scatter(x_nd,[0]*len(x_nd),c='k', s=5, label="nodes")
```

```
17  ax.set_xlabel('x')
18  ax.set_title(f'Nodal shape funcions')
19  ax.grid(color='r', linestyle=':', linewidth=0.2)
20  ax.axvline(x=0, c="k", lw=0.6); ax.axhline(y=0, c="k", lw=0.4)
21
22  ax.legend(loc='center', bbox_to_anchor=(0.8, 0.3))
23  plt.xlim(1, 7)
24  plt.savefig('images/nodal_N.png', dpi=500)
25  plt.show()
```

Nodes: [1 2 3 4 5 6 7 8 9 10 11]

We note the following from Fig. 5.8:

It is seen that the shape function for Node 1 on the left boundary has only one straight line (blue) on its right. Node 7 on the right boundary has only one straight line (pink) on its left. The shape functions of all the other nodes in the interior of the domain have a triangle shape. At any point of x, the sum of the value of the straight lines give unity. This set of nodal shape functions shown in Fig. 5.8 are all linear, but one can create higher orders in the similar way using the Lagrange interpolators.

5.10 Remarks

1. Functions can form vector spaces. A vector space is spanned using the basis functions.
2. The basic condition for a set of functions being basis functions is that all the functions in the set are linearly independent. All the types of basis functions discussed in the chapter satisfy this condition.

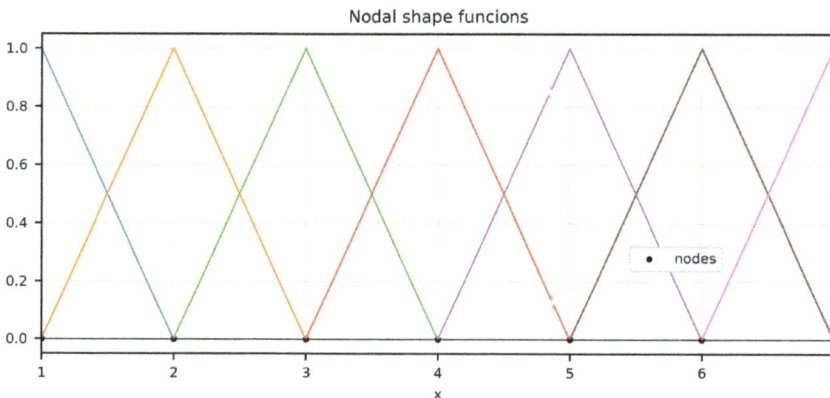

Figure 5.8. Nodal shape functions created locally each for a node using the Lagrange interpolators.

3. Linear independence can be achieved by letting each basis function have a different power or letting the basis functions have different home nodes. In the former cases, the order of the basis functions are different. In the latter cases, the order of the basis functions can be the same.
4. Linear independence can be better achieved through piecewise construction of the shape functions, so that each basis function has its own home node.

Chapter 6 presents the applications of all these basis functions to approximate various types of functions.

References

[1]   Liu GR and Quek SS, *The Finite Element Method: A Practical Course*, Butterworth-Heinemann, 2013.
[2]   Liu GR, *Mesh Free Methods: Moving Beyond the Finite Element Method*, Taylor and Francis Group, New York, 2010.
[3]   Liu GR and Nguyen TT, *Smoothed Finite Element Methods*, Taylor and Francis Group, New York, 2010.

Chapter 6

Function Approximation

Contents

We import the following Python modules for later use in this chapter:

```python
import sys
sys.path.append('../grbin/')                    # for the use of our own module
from commonImports import *                     # often used external modules
import grcodes as gr
importlib.reload(gr)                  # reload when changes are made to grcodes
#plt.rcParams.update({'font.size': 6})          # control plot font size
#np.set_printoptions(precision=4, suppress=True)          # print digits
```

In computational methods, one of the most important tasks is to find unknown functions that best describe problems in science or engineering. To achieve such a purpose, we often need to approximate the function in some way. This is true especially for complicated problems in real life because the solutions are unlikely in the form of elementary functions discussed in Chapter 4. There are quite a number of ways for such an approximation, as summarized in Chapter 2 of Ref. [5]. The most widely used one is the so-called series representation method. It expresses an unknown function as a linear combination of a special set of known functions called basis functions.

As discussed in the previous chapter, basis functions are linearly independent and span a function vector space. Any function in the space can be expressed as a linear combination of the basis functions. Therefore, if the unknown function lives in that space, the approximated function will be exact because of the closure property of a vector space.

Conversely, if the unknown function does not live in the space that the basis functions span, such a linear combination of these basis functions gives an approximation. In this case, the accuracy of the approximation depends on (a) the complexity and continuity properties of the function to be approximated and (b) the property of the basis functions used.

We have discussed a number of qualified types of basis functions in the previous chapter. This chapter presents the techniques with Python code demonstrations for approximating functions using these basis functions.

6.1 Use of monomial basis functions

The first and simplest set of basis functions is the monomials discussed in detail in the previous chapter. A set of $(n+1)$ monomial basis functions are given as

$$1, x^1, x^2, \ldots, x^n \tag{6.1}$$

This set of monomials is distinct, linearly independent, and is not orthogonal. It is a complete set of basis functions up to degree of n because it contains all these monomials with degrees lower than or equal to n. There is no missing monomials before x^n in Eq. (6.1). Such a set of basis functions forms a function space of all polynomial functions up to degree n.

Note that we always use complete basis functions unless there is a special reason for not doing so. If we know the function to be approximated does not contain, or can be better approximated without, some order terms, we can exclude these terms from the basis. The most often-encountered case is when the function to be approximated is an even (or odd) function, we can exclude all the monomials with even (or odd) orders.

6.1.1 *Formulation*

In general, when a function is approximated, it is done in a prescribed domain. Let it be $[a, b]$. The approximation needs some information about the function, typically its values at discrete points. Consider $(m + 1)$ points called nodes: $[a = x_0, x_1, \ldots, b = x_m]$. Assume that the values of a function $f(x)$ are known at these nodes as $[f(x_0), f(x_1), \ldots, f(x_m)]$. The function value at any point $x \in [a, b]$ can be approximated in a linear combination of monomials:

$$\langle f(x) \rangle = c_0 x^0 + c_1 x^1 + c_2 x^2 + \cdots + c_n x^n = \sum_{i=0}^{n} c_i x^i \tag{6.2}$$

where $\langle \cdot \rangle$ denotes $\cdot$ is approximated and c_i are $(n + 1)$ unknown scalar coefficients. Since we know the function values at these $(m + 1)$ nodes, using Eq. (6.2), we obtain the following:

$$
\begin{aligned}
f(x_0) &= c_0 x_0^0 + c_1 x_0^1 + c_2 x_0^2 + \cdots + c_n x_0^n \\
f(x_1) &= c_0 x_1^0 + c_1 x_1^1 + c_2 x_1^2 + \cdots + c_n x_1^n \\
&\vdots \\
f(x_m) &= c_0 x_m^0 + c_1 x_m^1 + c_2 x_m^2 + \cdots + c_n x_m^n
\end{aligned}
\tag{6.3}
$$

This gives $(m+1)$ equations called a system of equations, or an equation system. We can then solve the equation system for $(n+1)$ unknown coefficients c_i. The resultant system matrix is known as the Vandermonde matrix. The solution procedure depends on m against n:

1. When $m = n$, Eq. (6.3) can be solved for c_i using the standard linear algebraic system equation solver. In this case, the solution of c_i will be unique because the Vandermonde matrix is invertible if the nodes are distinct (for 1D domains).
2. When $m > n$, Eq. (6.3) can be solved for c_i using the standard least-square system equation solver. The solution of c_i ensures the sum of the approximation errors squared is minimized. This technique is known as least-squares fitting.
3. When $m < n$, Eq. (6.3) is underdetermined and the solution of c_i will not be unique. An infinite set of solutions can be found. One may also obtain solutions in a minimum-length sense [4]. A more practical approach may be to reduce the number of basis functions by removing the higher order ones, so that $m = n$ or $m > n$.

In this book, we will not discuss how such an equation system is solved because it is a quite intensive topic by itself. Questions on solution existence, solution procedures, column and null spaces, and efficiency to get the solution need to be answered. Hence, it requires a separate volume for sufficient in-depth discussion. Here, we will simply use existing Python code functions to find the solutions. Also, our study focuses on the first two cases, $m = n$ and $m > n$, for meaningful solutions of c_i.

After coefficients c_i are found, we substitute c_i into Eq. (6.2), which gives the approximated function.

The process mentioned above is generally called curve fitting and can be done using polyfit() in Numpy.

6.1.2 *Python examples: Approximating a polynomial*

We now write some Python codes to generate monomial basis functions and use these to approximate some functions. These functions will be plotted with horizontal axis for variable x values and vertical axis for the corresponding function values.

The process consists of the following steps:

1. Create a true function that is a polynomial with degree of 3.
2. Use polyfit() to obtain the approximated (fitted) function.

3. Plot both the approximated and true functions.

4. Compute the error of the approximated function against the true one in terms of the rooted mean squared error (rmse). The rmse is defined as the square root of the averaged squared differences between the approximated function and the true function at these nodes.

We first write a code function for performing the fitting. The code is self-explanatory with comment lines. This code function will be used multiple times in this chapter.

```python
def PolyAprox(f_true, Ns, x_nodes, Pfit, Px):
    '''To plot the curve fitted using a set of basis functions, and
    with the true function to be approximated.

    f_true:  numpy function, the true function
    Pfit:    numpy polynomial fitting object for fitting
    Px:      numpy polynomial object
    Ns:      list, degrees of approximation
    x_nodes: array like, nodes on x-axis used for fitting
    '''
    fx_nodes = [f_true(xi) for xi in x_nodes]  # true f vals for fitting

    for i in range(len(Ns)):
        P_coef = Pfit(x_nodes, fx_nodes, Ns[i])      # fit f for coeffs
        P_ap_x = Px(P_coef)            # compute the fitted function of x

    return P_ap_x
```

Now, we use the aforementioned code function to fit some functions, first, a polynomial:

```python
import numpy.polynomial.polynomial as poly

# Define the true function
f = lambda x: 2.1+5.7*x**2+1.5*x**3         # deg.3 poly to be approximated

a, b = -1.8, 1.8                                      # Interval
Ns = [3]                                     # degrees of approximation

x_nodes = np.linspace(a, b, Ns[-1]+1)    # Evenly spaced nodes on x-axis
                   # for lower orders (m>n), least square fitting is used

Pfit = poly.polyfit                      # fit to obtain the coefficients
P_type = 'Polynomial'                           # Use monomials
Px = poly.Polynomial                            # Create an object

P_ap_x = PolyAprox(f, Ns, x_nodes, Pfit, Px)
```

```
18  P_ap_x.coef = [float(f"{num:.4f}") for num in P_ap_x.coef]
19
20  print(f'Coefficients of {P_type}: {P_ap_x.coef}\n')
21  print(f'Approximated function of {P_type}:')
22  P_ap_x
```

Coefficients of Polynomial: [2.1, 0.0, 5.7, 1.5]

Approximated function of Polynomial:

$$x \mapsto 2.1 + 0.0\,x + 5.7\,x^2 + 1.5\,x^3$$

The fitting of a polynomial using basis polynomials with the same degree gives back the same polynomial. This is because the polynomial lives in the space that is spanned by the monomials of sufficient complete order.

Let us now try to approximate a non-polynomial function. In this case, it is not living in the space spanned by the monomials. Thus, we can only expect to obtain an approximated solution.

```
1  # Define the true function
2  f = lambda x: 0.2*x**3+x*np.sin(3*x)        # Non-polynomial to approximate
3  Ns = [3]                                     # Highest degree of approximation
4
5  x_nodes = np.linspace(a, b, Ns[-1]+1)        # Evenly spaced nodes on x-axis
6                              # for lower orders, least square fitting is used
7
8  Pfit = poly.polyfit                          # fit to obtain the coefficients
9  P_type = 'Polynomial'                              # Use monomials
10 Px = poly.Polynomial                              # Create an object
11
12 P_ap_x = PolyAprox(f, Ns, x_nodes, Pfit, Px)
13
14 P_ap_x.coef = [float(f"{num:.4f}") for num in P_ap_x.coef]
15
16 print(f'Coefficients of {P_type}: {P_ap_x.coef}\n')
17 print(f'Function of {P_type}:')
18 P_ap_x
```

Coefficients of Polynomial: [0.8312, 0.0, -0.6859, 0.2]

Function of Polynomial:

$$x \mapsto 0.8312 + 0.0\,x - 0.6859\,x^2 + 0.2\,x^3$$

The solution obtained is a polynomial presentation of the original non-polynomial function.

Next, let us write a code function for performing the fitting, plotting these functions and errors. This time, we use different order of basis polynomials to do the task. The code is self-explanatory with comment lines. This code function will also be used multiple times in this chapter.

```python
def plot_PolyAprox(f_true, Ns, x_nodes, X, Pfit,Px,P_type='Chebyshev'):

    '''To plot the curve fitted by a set of basis functions, together
    with the original function to be approximated.

    f_true:  numpy function, the true function
    Pfit:    numpy polynomial fitting object
    Px:      numpy polynomial object
    Ns:      list, degrees of approximation
    x_nodes: array like, nodes on x-axis used for fitting
    X:       array like, points on x-axis for plots
    P_type:  string, type of basis polynomial used, for file name.
    '''
    #fX = [f_true(xi) for xi in X]           # true function for plots
    fX = f_true(X)                           # true function for plots

    #fx_nodes = [f_true(xi) for xi in x_nodes] # true f vals for fitting
    fx_nodes = f_true(x_nodes)               # true f vals for fitting

    plt.rcParams.update({'font.size': 6})
    fig_s = plt.figure(figsize=(6,5))

    for i in range(len(Ns)):
        ax = fig_s.add_subplot(2,2,i+1)
        P_coef = Pfit(x_nodes, fx_nodes, Ns[i])      # fit f for coeffs
        P_ap_x = Px(P_coef)                          # fitted polynomial
        P_apX  = P_ap_x(X)                  # Compute the fitted results at X
                                                     # for plotting
        ax.plot(X,P_apX,'mo',ms=2,label='Poly deg.'+str(Ns[i])+
                f', rmse: {sqrt(np.mean((P_apX-fX)**2)):.4e}')
        ax.plot(X, fX, c='k', label='Original Function')

        if i==2 or i==3: ax.set_xlabel('x')
        if i==0 or i==2: ax.set_ylabel('$f(x)$')
        ax.legend()        #loc='center right', bbox_to_anchor=(1, 0.5))

    plt.savefig('images/'+P_type+'.png',dpi=500,bbox_inches='tight')
    plt.show()

    return P_ap_x
```

Let us use our code function to compute and plot the results for fitting the polynomial:

```python
1  import numpy.polynomial.polynomial as poly
2  # Define the true function
3  f = lambda x: 2.1+5.7*x**2+1.5*x**3        # deg.3 poly to be approximated
4
5  a, b = -1.8, 1.8                                      # Interval
6  Ns = [1, 2, 3, 4]                           # degrees of approximation
7
8  x_nodes = np.linspace(a, b, Ns[-1]+1)      # Evenly spaced nodes on x-axis
9                             # for lower orders, least square fitting is used
10
11 X = np.linspace(a, b, 100)                    # points on x-axis for plots
12
13 Pfit = poly.polyfit                        # fit to obtain the coefficients
14 P_type = 'Polynomial'                             # Use monomials
15 Px = poly.Polynomial                            # Create an object
16
17 P_ap_x=plot_PolyAprox(f, Ns, x_nodes, X, Pfit, Px, P_type=P_type+'3rdP')
18 P_ap_x.coef = [float(f"{num:.4f}") for num in P_ap_x.coef]
19 P_ap_x
```

$$x \mapsto 2.1 + 0.0\,x + 5.7\,x^2 + 1.5\,x^3 + 0.0\,x^4$$

For this polynomial of degree 3, we found the following from Fig. 6.1:

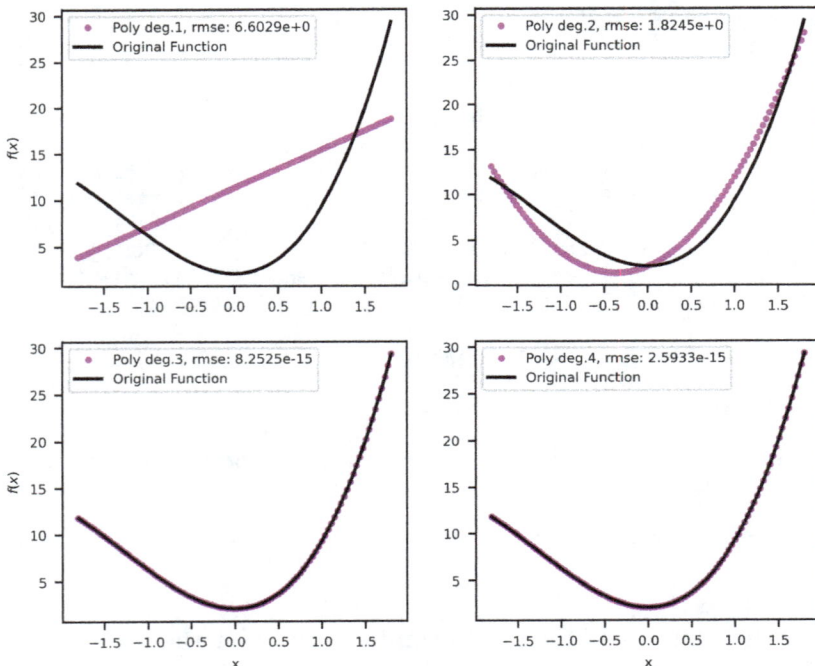

Figure 6.1. Approximated third-order polynomial using monomials with different degrees.

1. The approximation error is high if the degree of (complete) basis monomials is less than 3. The approximation is least-square fitting, which gives a kind of overall representation.
2. When the degree of basis monomials is 3, the error is zero (to machine accuracy). The approximated function is a perfect fit to the true function values.
3. When the degree of basis monomials is 4, the error is also zero, implying that one can use a higher degree to approximate, but not less. This is because any polynomial of degree 3 lives in the space formed by the basis monomials with completed degree 4. The use of too higher a degree, however, would increase the computational cost.

6.1.3 *Python examples: Approximating an arbitrary function*

Let us now use the same code function to approximate the non-polynomial function studied earlier:

```python
# Define the true function
f = lambda x: 0.2*x**3+x*np.sin(3*x)      # Non-polynomial to approximate
Ns = [2, 4, 6, 8]                         # Highest degree of approximation

x_nodes = np.linspace(a, b, Ns[-1]+1)     # Evenly spaced nodes on x-axis
X = np.linspace(a, b, 100)                # points on x-axis for plots

P_ap_x=plot_PolyAprox(f,Ns,x_nodes,X, Pfit, Px,P_type=P_type+'Non_Poly')
P_ap_x.coef = [float(f"{num:.3f}") for num in P_ap_x.coef]
P_ap_x
```

$$x \mapsto 0.0 - 0.0\,x + 2.968\,x^2 + 0.2\,x^3 - 4.273\,x^4 - 0.0\,x^5 + 1.635\,x^6 - 0.0\,x^7 - 0.197\,x^8$$

For this non-polynomial function, we found the following from Fig. 6.2:

The approximation error is obvious, but the error reduces with the increase of the order of the basis functions. The resulting polynomial is the least-square fit of the non-polynomial true function.

Note that these coordinate points x_i do not have to be evenly placed when the domain is 1D. The following code uses randomly generated points x_i and sorts these in ascending order. To ensure that the two boundary points are included, we insert and append these to the randomly generated arrays:

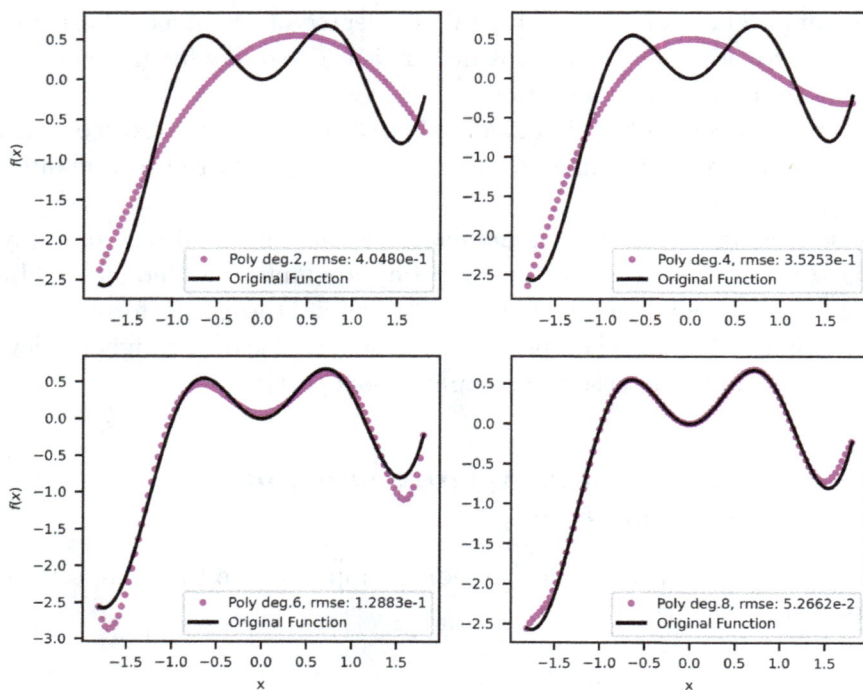

Figure 6.2. Approximated a non-polynomial function using monomials with different degrees.

```python
1   import numpy.polynomial.polynomial as poly
2   np.set_printoptions(precision=4, suppress=True)          # print digits
3   np.random.seed(8)
4   # Define the true function
5   f = lambda x: 2.1+5.7*x**2+1.5*x**3        # deg.3 poly to be approximated
6
7   a, b = -1.8, 1.8                                              # Interval
8   n = 4                                        # degree of the fitting function
9   x_nodes = np.sort(np.random.uniform(a, b, n+1))         # random points
10  x_nodes = np.append(np.insert(x_nodes, 0, a), b)        # boundary points
11  print(x_nodes)
12
13  Ns = [1, 2, 3, 4]                                      # degrees for fitting
14  P_type = 'Polynomial'                                      # Use monomials
15  Px = poly.Polynomial                                      # Create an object
16  Pfit = poly.polyfit                               # fit to obtain the coefficients
17
18  X = np.linspace(a, b, 100)                   # points on x-axis for plots
19
20  plot_PolyAprox(f, Ns, x_nodes, X, Pfit, Px, P_type=P_type+'3rdPr')
```

```
[-1.8    -0.9622  0.1111  1.3291  1.3443  1.6867  1.8    ]
```

$$x \mapsto 2.0999999999999974 + 0.0\,x + 5.700000000000001\,x^2 + 1.499999999999$$
$$9991\,x^3 - 1.0137156680385488\text{e-}16\,x^4$$

From Fig. 6.3, we observed the same findings for approximating the polynomial function using randomly spaced coordinates. The following is for the non-polynomial function:

```python
1   # Define the true function
2   f = lambda x: 0.2*x**3+x*np.sin(3*x)     # Non-polynomial to approximate
3   Ns = [2, 4, 6, 8]                        # Highest degree of approximation
4
5   x_nodes = np.sort(np.random.uniform(a, b, Ns[-1]+1))     # random nodes
6   x_nodes = np.append(np.insert(x_nodes, 0, a), b)   # add boundary points
7   print(x_nodes)
8
9   X = np.linspace(a, b, 100)                    # points on x-axis for plots
10
11  plot_PolyAprox(f, Ns, x_nodes, X, Pfit, Px, P_type=P_type+'Non_Polyr')
```

```
[-1.8     -1.759  -0.3515 -0.2503 -0.0778  0.0816  0.1562  0.1993  0.7645
  0.9392   1.8    ]
```

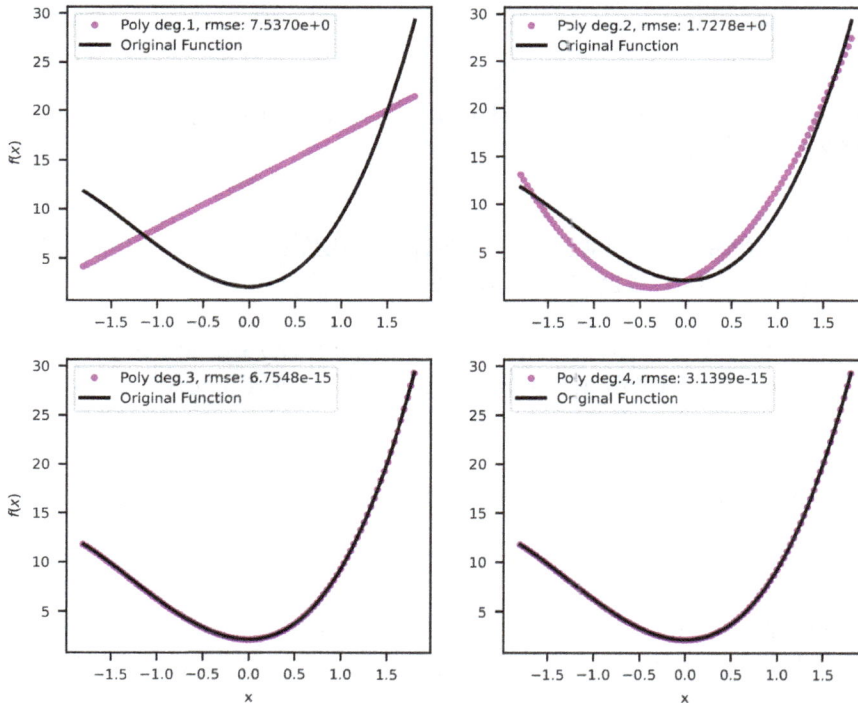

Figure 6.3. Approximated a third order polynomial function using monomials with different degrees and randomly sampled function values.

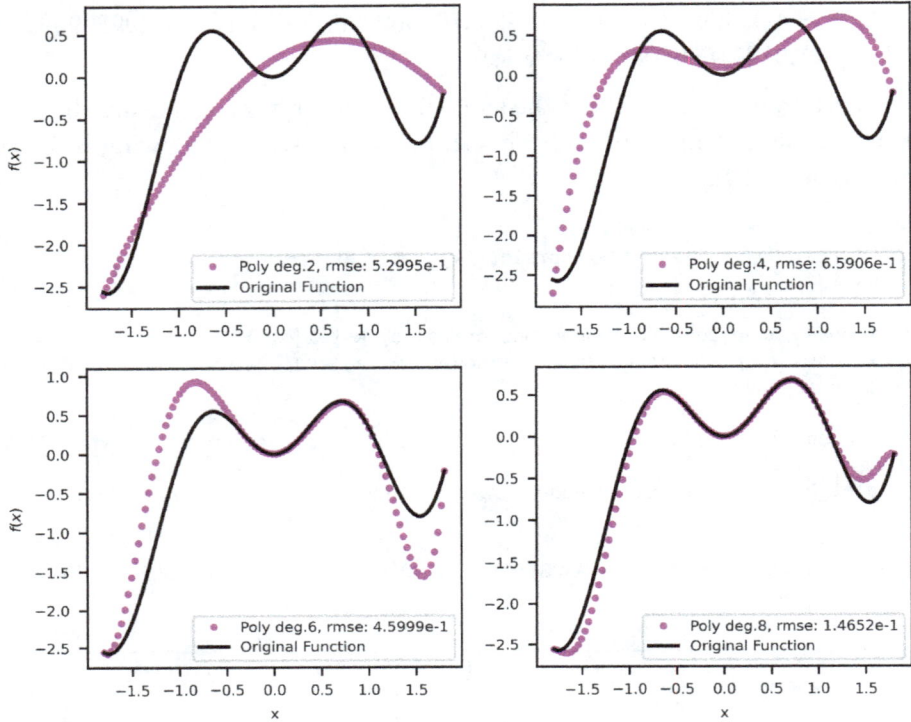

Figure 6.4.   Approximated a non-polynomial function using monomials with different degrees and randomly sampled function values.

$$x \mapsto 1.2831\text{e-}05 + 0.0006436685\,x + 2.9973\,x^2 + 0.17891\,x^3 - 4.4481\,x^4 + 0.13193\,x^5 + 1.7337\,x^6 - 0.03872987\,x^7 - 0.21211\,x^8$$

From Fig. 6.4, the error found using randomly sampled function values is higher compared to that found using evenly spaced coordinates.

6.2   Use of Lagrange polynomials

6.2.1   *Formulation*

Consider a domain $[a, b]$ (in the finite element method (FEM) [7], it is called element) that is discretized into n intervals with $(n + 1)$ distinct nodes: $[a = x_0, x_1, \ldots, b = x_n]$. If the values of a function $f(x)$ are assumed to be known at all these nodes as $[f(x_0), f(x_1), \ldots, f(x_n)]$, the function value at any point x within the domain can be interpolated using

$$f(x) = \sum_{i=0}^{n} f(x_i) l_i(x) \qquad (6.4)$$

where l_i is a Lagrange basis function (interpolator) corresponding to node i given in Eq. (5.8). It is a type of **shape function** used in the FEM, often denoted as $N_i(x)$.

6.2.2 *Overfitting phenomenon*

Since the degree of Lagrange interpolation is directly related to the number of nodes used in the domain $[a, b]$, one can end up with using polynomials of too high a degree if the domain is too big and too many nodes are needed to capture the variations of the function. Since it is a node-passing interpolation, using a high degree can often lead to the so-called overfitting problem [3]. In such a case, the function value at these nodes are exactly reproduced, but the values of the approximated function can oscillate drastically between the nodes, especially near the domain boundary and outside the domain. Therefore, such an approximation is unlikely to have a good performance for points in these locations. To overcome this problem, domain division may be used.

Domain division: One way to overcome this is to divide the domain into smaller sub-domains (elements), resulting in a mesh with nodes. Lagrange interpolations are then used in each of these sub-domains, known as piecewise interpolation or approximation. This can cut down the number of nodes in a sub-domain effectively, and shape functions can be created for each of the nodes for function approximations (to be discussed in detail in the last section in this chapter). However, the smoothness of the connection between these sub-domains can be in question. The Lagrange interpolators are continuous at the inter-points of these sub-domains, and the derivatives (slopes) are usually not. This is known as C^0 continuity. Therefore, when piecewise Lagrange interpolations are used for approximating a function in a domain, special care is needed. Often, for stable approximations, we would need to resort to the so-called weak or weakened weak formulations. This is precisely what we do in the finite element method [7], smoothed finite element method [6], and meshfree methods [2].

The alternatives for resolving the overfitting interpolation problem include using a moving least approximation [2], implying the use of lower-order basis functions. Other techniques are smoothed particle approximation [5] and a B-spline (or Nurbs). These techniques are also often used together with a mesh.

6.2.3 *Python example: Approximating a polynomial*

We first write a code to perform interpolations for a set of given discrete values of a function using the Lagrange interpolators. The following code is self-explanatory:

```python
 1  def lagrange_interplation(order, a, b, f, X):
 2
 3      '''compute interpolated values for given np.array X, using the
 4      Lagrange interpolators for given nodes in an interval [a, b].
 5      input:
 6              order -- the order of the interpolation.
 7      return: f_interp_X -- np.arary of X.shape
 8              Ns_np -- list, with numpy shape functions
 9              Ns_sp -- list, with sympy shape functions
10      '''
11      x = sp.symbols('x')                         # define a symbolic variable
12      dx = (b-a)/order                              # nodal spacing
13      Nodes = [a+n*dx for n in range(order+1)]      # generate nodes
14
15      Ns_sp  = gr.LagrangeP(x, Nodes)          # List: shape functions
16      Ns_np = [lambdify(x, fs, "numpy") for fs in Ns_sp]
17
18      f_interp_X = np.zeros(X.shape) # perform the interpolation in numpy
19      for n in range(len(Nodes)):
20          f_interp_X += f(Nodes[n])*Ns_np[n](X)
21
22      return Nodes, f_interp_X, Ns_np, Ns_sp
```

The following code performs interpolation using the Lagrange interpolators with up to fourth order and plots these curves together with the curve of the true function:

```python
 1  def plot_LgrngAprox(p_name='tmp'):
 2
 3      '''To plot the curve fitted by a set of Lagrange basis functions,
 4      together with the original function to be approximated.
 5      '''
 6      Labels=[str(n)+' deg' for n in degrees] # ['1st degree','2nd ',...]
 7
 8      plt.rcParams.update({'font.size': 5})          # settings for plots
 9      fig_s = plt.figure(figsize=(4,2.5))
10      ax = fig_s.add_subplot(1,1,1)
11
12      ax.plot(X, f(X), c='k', label='f(x)')          # plot the true curve
13
14      for i, n in enumerate(degrees):                # plot interpolated ones
15          Nodes, fX, _, _ = lagrange_interplation(n, a, b, f, X)
16          ax.plot(X, fX, 'o', ms=2, label=Labels[i]+
17                  f', rmse: {sqrt(np.mean((fX-f(X))**2)):.4e}')
18
```

```
19      ax.scatter(Nodes,[0]*len(Nodes),c='k',s=7,label='nodes for '+Labels [-1])
20      ax.grid(c='r', linestyle=':', linewidth=0.5)
21      ax.axvline(x=0, c="k", lw=0.6)
22      ax.axhline(y=0, c="k", lw=0.6)
23      ax.set_xlabel('x')
24      ax.legend() #loc='center right', bbox_to_anchor=(1, 0.5))
25      plt.title('Approximation '+p_name+' using Lagrange interpolators')
26
27      plt.savefig('images/Lgrng'+p_name+'.png',dpi=500,bbox_inches='tight')
28      plt.show()
```

Now, we first define a true function that is a third-order polynomial, perform the interpolation, and then plot the interpolated function together with the true function:

```
1  # define a function
2  def f(x):
3      return 9*x**3 - 5*x**2 -3*x + 2 #np.cos(np.pi*x) # the true function
4
5  a=-1; b=1                       # define the domain for approximation
6  X = np.linspace(a, b, 50)               # range in real axis for plots
7
8  # define various degrees for generating Lagrange interpolators
9  degrees = [1, 2, 3, 4]
10 plot_LgrngAprox(p_name='3_deg_Poly')
```

From Fig. 6.5, we found the following:

1. When the order of Lagrange interpolators is lower than 3 (the linear and second-order ones), the interpolated curves are far off.

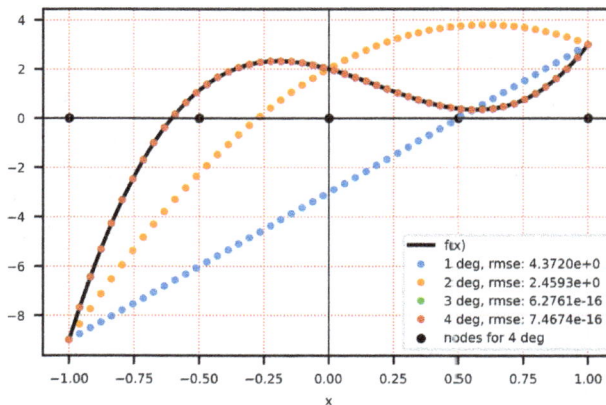

Figure 6.5. Approximated a third-order polynomial function using the Lagrange interpolators with different degrees.

2. The Lagrange interpolator is a node-passing approximation, as defined in Eq. (6.4). For example, when one degree is used, the approximated function is a straight line (blue) passing through the two boundary points. This implies that the approximation directly captures the lower-order features, and the nodal values are exactly reproduced. It is quite different from the least-square approximation discussed in the previous section.

3. When the order is higher than or equal to 3, the third-order true polynomial is exactly produced (to machine accuracy).

When the curve is not a polynomial, the interpolated curve can only be an approximation. The following section describes an example for a sine function.

6.2.4 *Python example: Approximating an arbitrary function*

We first define a true function that is a sine and then plot the approximation results using the same code function:

```python
def f(x):
    return np.sin(1.2*np.pi*x)          # the true function

degrees = [1, 3, 5]
plot_LgrngAprox(p_name='sine_function')          # plot the results
```

From Fig. 6.6, we found the following for this sine function:

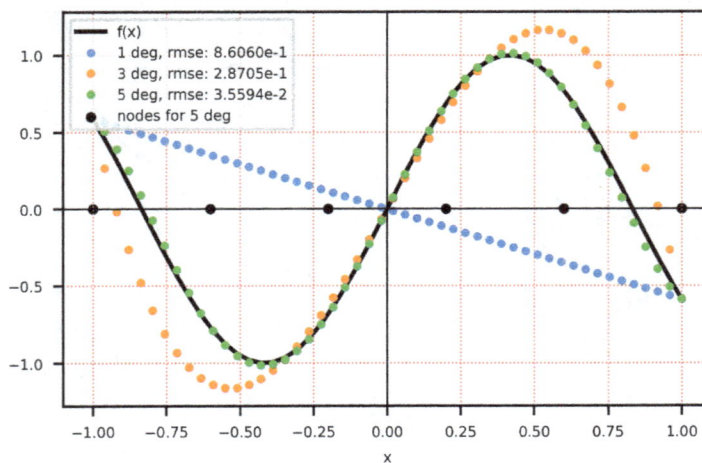

Figure 6.6. Approximated a sine function using the Lagrange interpolators with different degrees.

1. The interpolated curves are far off when the order of Lagrange basis functions (interpolators) is first- and third-order ones. The approximation always passes the nodes.
2. When the order is increased, the interpolated curve approaches closer to the true function. When fifth order is used, the interpolated curve becomes quite close to the true function.

6.2.5 *Sympy built-in Lagrange interpolate*

Sympy has a built-in Lagrange interpolate() method for the interpolation of functions. However, the interpolators are not accessible. The following is an example:

```
1  from sympy.polys.polyfuncs import interpolate
2
3  points = [(0, 1), (1, 3), (2, 5)]     # Define some points to interpolate
4
5  x = sp.symbols('x')                   # Define the variable to interpolate at
6
7  poly = interpolate(points, x)         # Compute the Lagrange polynomial
8
9  print(poly)                           # Print the interpolating polynomial
10
11 # Evaluate the polynomial at a particular value of x
12 print(poly.evalf(subs={x: 1.5}))
```

```
2*x + 1
4.00000000000000
```

6.3 Use of Chebyshev polynomials

6.3.1 *Formulation*

Chebyshev polynomials also form the basis functions for approximating a function. They are given in Chapter 3, and the first kind $T_n(x)$ defined by its domain $[-1, 1]$ is rewritten here:

$$T_0(x) = 1 \qquad\qquad \text{for } n = 0$$
$$T_1(x) = x \qquad\qquad \text{for } n = 1 \qquad\qquad (6.5)$$
$$T_n(x) = 2x\, T_{n-1}(x) - T_{n-2}(x) \qquad \text{for } n > 1$$

There are a couple of ways to use Chebyshev polynomials as the basis functions for approximating functions. Here, we introduce a primitive method.

Unlike the Lagrange interpolation discussed in the previous section, Chebyshev basis functions do not have the Delta function property, and hence, we cannot use Eq. (6.4). We need to "fit" the function using the Chebyshev basis functions to compute the coefficients. The process is essentially the same as using the monomial basis discussed in Section 6.1.1.

Consider a function $f(x)$ defined over a physical domain $[a, b]$. It is approximated using

$$\langle f(x) \rangle = c_0 T_0(x) + c_1 T_1(x) + c_2 T_2(x) + \cdots + c_n T_n(x) = \sum_{k=0}^{n} c_k T_k(x) \quad (6.6)$$

where c_k are $(n+1)$ unknown scalar coefficients known as the Chebyshev coefficients and $T_k(x)$ are the Chebyshev polynomials of the first kind. Compared to Eq. (6.2), the difference is the basis functions.

However, because the Chebyshev basis is defined in the standard domain $[-1, 1]$ to approximate function $f(x)$ defined in a physical domain $[a, b]$, we need to perform linear mapping between $T_k(x)$'s domain $[-1, 1]$ and $f(x)$'s domain $[a, b]$. We obtain

$$\langle f(x) \rangle = \sum_{k=0}^{n} c_k T_k \underbrace{\left(\frac{2x - a - b}{b - a} \right)}_{\xi} = \sum_{k=0}^{n} c_k T_k(\xi) \quad (6.7)$$

where ξ is the natural coordinate in $[-1, 1]$. It relates to the physical coordinate x as

$$\xi = \frac{2x - a - b}{b - a} \quad (6.8)$$

It is seen when $x = a$, T_k is at $\xi = -1$, and when $x = b$, T_k is at $\xi = +1$. When mapping back to the physic domain, we shall have

$$x = \frac{\xi(b - a) + a + b}{2} \quad (6.9)$$

As seen in Eqs. (6.8) and (6.9), the coordinate mapping involves both shifting and scaling.

Assume that we know the function values at $(m + 1)$ locations, $[f(x_0), f(x_1), \ldots, f(x_m)]$, using Eq. (6.7), we obtain the following:

$$f(x_0) = c_0 T_0(\xi_0) + c_1 T_1(\xi_0) + c_2 T_2(\xi_0) + \cdots + c_n T_n(\xi_0)$$

$$f(x_1) = c_0 T_0(\xi_1) + c_1 T_1(\xi_1) + c_2 T_2(\xi_1) + \cdots + c_n T_n(\xi_1)$$

$$\vdots \quad (6.10)$$

$$f(x_m) = c_0 T_0(\xi_m) + c_1 T_1(\xi_m) + c_2 T_2(\xi_m) + \cdots + c_n T_n(\xi_m)$$

This equation set has exactly the same form as Eq. (6.3). The only difference is that the monomial bases are replaced by the Chebyshev bases and an additional coordinate mapping. Since the Chebyshev basis are linearly independent (as discussed in the previous chapter), the methods used in computing the coefficients c_k are the same as solving Eq. (6.3), and we will not repeat them here.

Solving Eq. (6.10) for c_k, we shall finally obtain an approximated $f(x)$:

$$\langle f(x_k) \rangle = \sum_{k=0}^{n} c_k T_k \left(\frac{2x - a - b}{b - a} \right) \tag{6.11}$$

The following is an example of using a method in the Python library: chebfit(). It uses a least-square fit to function values at a set of discrete points, following largely the formulations given previously.

6.3.2 *Python example: Approximating a polynomial*

```
1  #help(Che.chebfit)
```

```
1  import numpy.polynomial as Poly
2  import numpy.polynomial.chebyshev as Che
3
4  f = lambda x: 2.1+5.7*x**2+1.5*x**3    # 3rd-deg. poly to be approximated
5  a, b = -1.8, 1.8                                        # Interval
6  Ns = [1, 2, 3, 4]                          # degrees used for approximation
7
8  x_nodes = np.linspace(a, b, Ns[-1]+1)      # Evenly spaced nodes on x-axis
9                                             # used for fitting
10 P_type = 'Chebyshev'
11 Px = Poly.Chebyshev
12 Pfit = Che.chebfit                 # use help(Che.chebfit) for details
13
14 X = np.linspace(a, b, 100)                 # points on x-axis for plots
15
16 plot_PolyAprox(f, Ns, x_nodes, X, Pfit, Px, P_type=P_type+'3rdP')
```

$x \mapsto 4.950000000000004\, T_0(x) + 1.1249999999999996\, T_1(x) + 2.85000000000 00005\, T_2(x) + 0.37500000000000017\, T_3(x) + 6.379735785916422\text{e-}17\, T_4(x)$

From Fig. 6.7, we found that for this third-order polynomial, the approximation error is high if the degree of Chebyshev polynomial is less than 3. When the degree of Chebyshev polynomial is 3, the error is zero (to machine accuracy). When the degree of Chebyshev polynomial is 4, the error is also zero, implying that one can use higher-order Chebyshev to approximate, but not less. This is because the polynomial to be approximated is within the

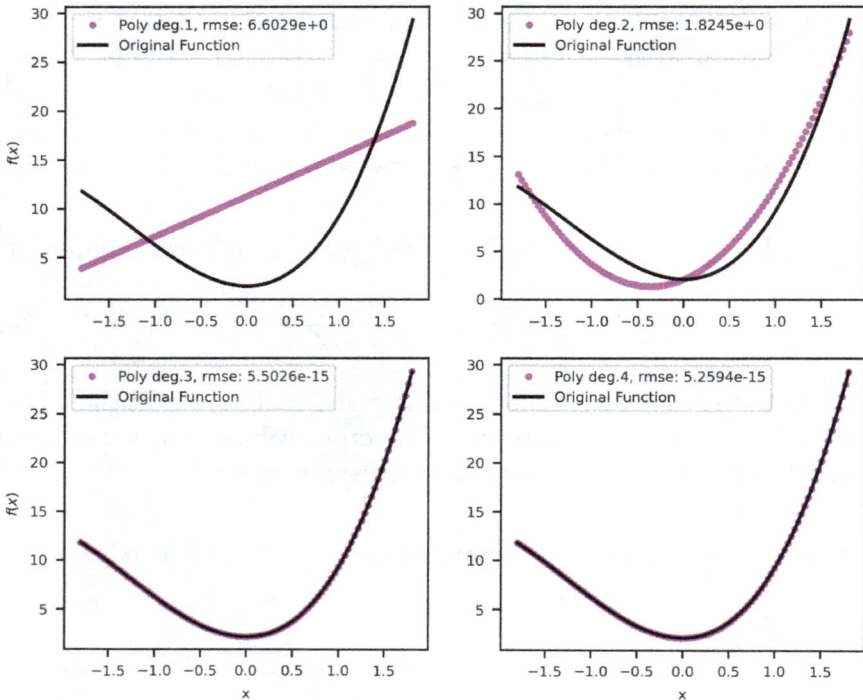

Figure 6.7. Approximated a 3rd order polynomial function using the Chebyshev polynomials with different degrees.

space of the Chebyshev polynomials. Of course, the use of too higher a degree would increase the computational cost.

6.3.3 *Python example: Approximating an arbitrary function*

Let us now try to approximate a general function rather than polynomials that the Chebyshev polynomial will not be able to reproduce. We do this first using evenly distributed nodes:

```
1  f = lambda x: 0.2*x**3 + x*np.sin(3*x)    # Non-polynomial to approximate
2  a, b = -1.8, 1.8                                              # Interval
3  n = 8                                          # degree of T_n(x) used for fitting
4  x_nodes = np.linspace(a, b, n+1)                 # Evenly spaced nodes on x-axis
5
6  Ns = [2, 4, 6, 8]                                  # degrees used for approximation
7
8  X = np.linspace(a, b, 100)                         # points on x-axis for plots
9
10 plot_PolyAprox(f,  Ns, x_nodes, X, Pfit, Px, P_type=P_type+'Non_Poly')
```

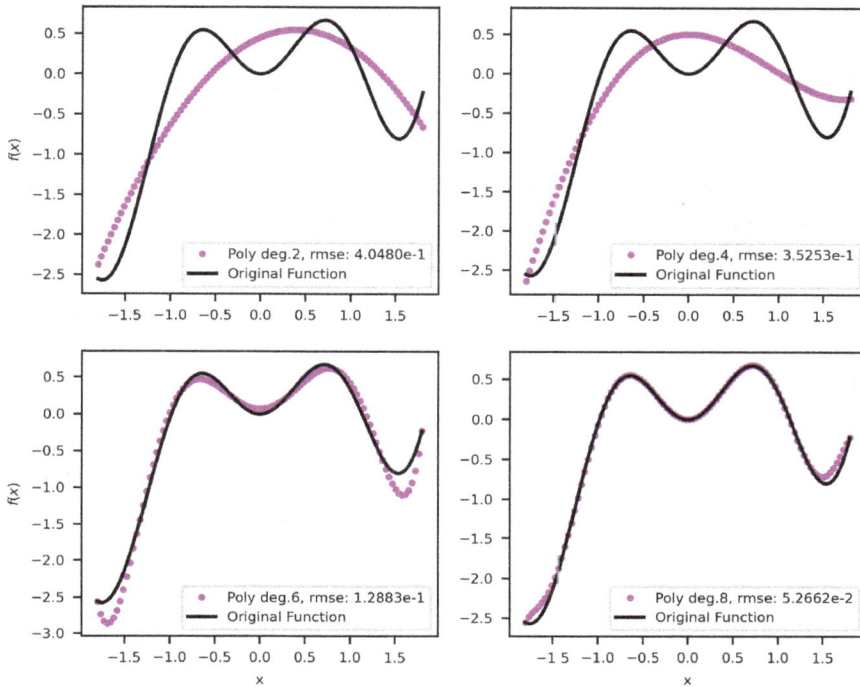

Figure 6.8. Approximated a non-polynomial function using the Chebyshev polynomials with different degrees.

$$x \mapsto 0.3386701787020543\, T_0(x) + 0.14999999999999866\, T_1(x) + 0.027626238$$
$$221115148\, T_2(x) + 0.04999999999999833\, T_3(x) - 0.2707534111272162\, T_4(x) +$$
$$7.979922050852875e\text{-}16\, T_5(x) + 0.03748010318209576\, T_6(x) - 5.5297283260$$
$$68205e\text{-}17\, T_7(x) - 0.0015425190355126185\, T_8(x)$$

As seen from Fig. 6.8, for this non-polynomial function, the Chebyshev can only approximate it because it is not in the polynomial function space. However, with the use of a higher degree of Chebyshev polynomials, the accuracy will increase. The convergence is fast for this smooth function (that is a composite of polynomial and sine function).

Next, we use randomly distributed nodes for the same task:

```
1  np.set_printoptions(precision=3, suppress=True)        # print digits
2
3  f = lambda x: 0.2*x**3 + x*np.sin(3*x)    # Non-polynomial to approximate
4  n, a, b =8, -1.8, 1.8
5
6  x_nodes = np.sort(np.random.uniform(a, b, n+9))
7  x_nodes = np.append(np.insert(x_nodes, 0, a), b)
8  print(x_nodes)
9
```

```
10  Ns = [2, 4, 6, 8]                          # degrees used for approximation
11
12  X = np.linspace(a, b, 100)                 # points on x-axis for plots
13
14  plot_PolyAprox(f, Ns, x_nodes, X, Pfit, Px, P_type=P_type+'Non_Polyr')
```

[-1.8   -1.759 -1.563 -1.012 -0.759 -0.598 -0.352 -0.266 -0.25   -0.078
  0.082 0.156 0.199 0.431 0.765 0.939 1.706 1.738 1.8   ]

$x \mapsto 0.33931663336668794\, T_0(x) + 0.1469543799994699\, T_1(x) + 0.032416236$
$49691194\, T_2(x) + 0.04863033882590761\, T_3(x) - 0.26477322369106165\, T_4(x) -$
$0.0008703705756676086\, T_5(x) + 0.03492691220725293\, T_6(x) + 9.72669842514$
$6111e\text{-}05\, T_7(x) - 0.0012464677195647494\, T_8(x)$

As seen from Fig. 6.9, when randomly sampled data are used, the error in approximating this non-polynomial function is a little higher, but still within the same order.

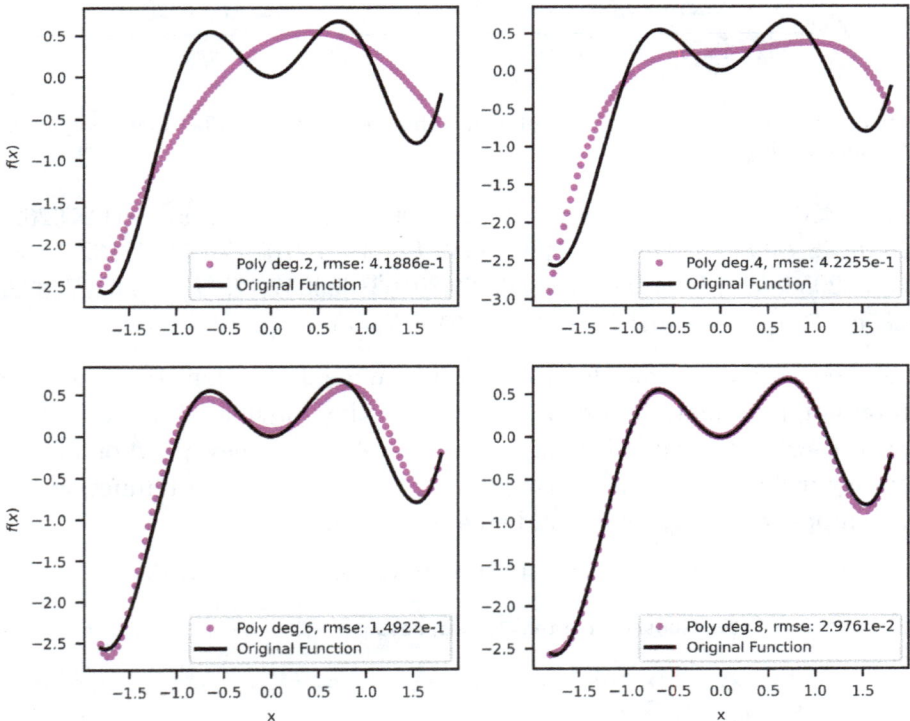

Figure 6.9. Approximated a non-polynomial function using the Chebyshev polynomials with different degrees and randomly sample function values.

6.4 Use of Legendre polynomials

6.4.1 *Formulation*

Legendre polynomials also form basis functions for approximating a function. This was mentioned in Chapter 3, defined in domain $[-1, 1]$, and is rewritten here:

$$P_0(x) = 1 \quad \text{for } n = 0$$

$$P_1(x) = x \quad \text{for } n = 1 \tag{6.12}$$

$$P_n(x) = \frac{1}{n}((2n - 1)xP_{n-1}(x) - (n - 1)P_{n-2}(x)) \quad \text{for } n > 1$$

The formulation for approximating a function using Legendre polynomials is essentially the same as that using the Chebyshev polynomial. The only difference is the change of basis. Thus, we will not repeat them, but present Python examples.

6.4.2 *Python example: Approximating a polynomial*

Numpy provides a legfit method to compute the coefficients of a Legendre series. It uses a least-square fit to function values at a set of nodes. We use the following code to demonstrate how it works:

```
1  #help(Lag.legfit)
```

```
1  import numpy.polynomial as Poly
2  import numpy.polynomial.legendre as Leg
3
4  f = lambda x: 2.1+5.7*x**2+1.5*x**3    # 3rd-deg. poly to be approximated
5  a, b = -1.8, 1.8                                      # Interval
6  Ns = [1, 2, 3, 4]                       # Highest degree of approximation
7
8  x_nodes = np.linspace(a, b, Ns[-1]+1)     # Evenly spaced nodes on x-axis
9
10 P_type = 'Legendre'                             # use Legendre basis
11 Px = Poly.Legendre
12 Pfit = Leg.legfit                       # use nelp(Leg.legfit) for details
13
14 X = np.linspace(a, b, 100)                   # points on x-axis for plots
15
16 plot_PolyAprox(f, Ns, x_nodes, X, Pfit, Px, P_type=P_type+'3rdP')
```

$x \mapsto 4.000000000000004 \, P_0(x) + 0.9000000000000036 \, P_1(x) + 3.80000000000 00016 \, P_2(x) + 0.5999999999999998 \, P_3(x) + 2.574399684806431\text{e-}16 \, P_4(x)$

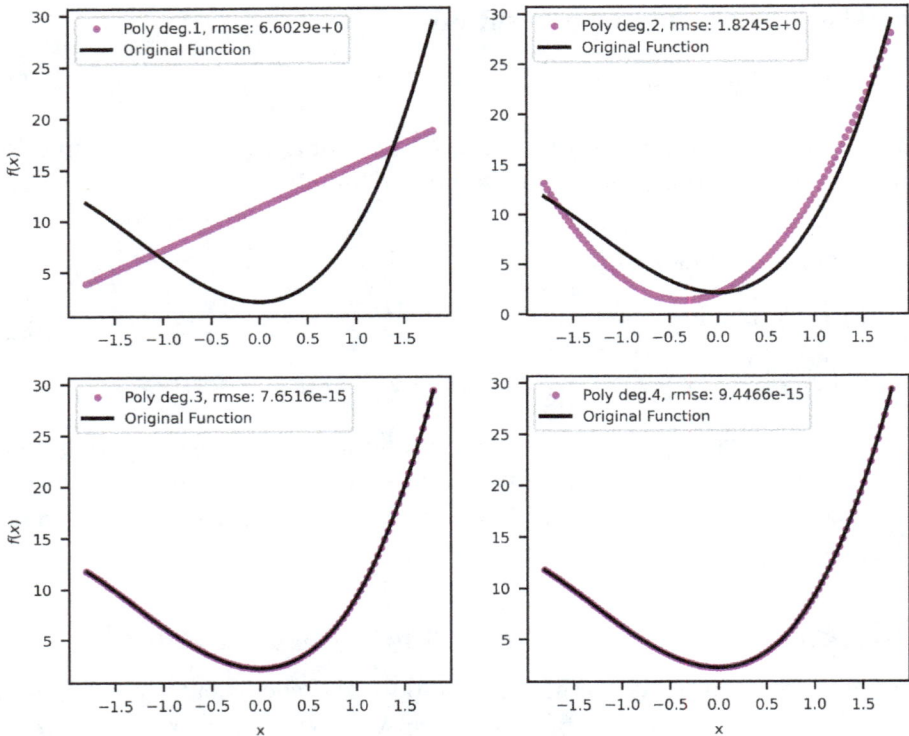

Figure 6.10. Approximated a third-order polynomial function using the Legendre polynomials with different degrees.

It is found from Fig. 6.10 that for this third-order polynomial, the approximation error is high if the degree of Legendre polynomial is less than 3. The solution is a least-square approximation. When the degree of Legendre polynomial is 3, the error is zero (to machine accuracy). When the degree of Legendre polynomial is 4 (at higher cost), the error is also zero, implying that one can use higher-order Legendre polynomials to approximate, but not less. This finding is similar to what we found for the Chebyshev polynomials. They are all essentially in the same polynomial space.

6.4.3 *Python example: Approximating an arbitrary function*

Let us now try to approximate a general function that the Legendre polynomial will not be able to reproduce.

```
1  f = lambda x: x**2 + x*np.sin(3*x)    # Non-polynomial to be approximated
2  a, b = -1.8, 1.8                                              # Interval
3  Ns = [2, 4, 6, 8]                            # Highest degree of approximation
4
5  x_nodes = np.linspace(a, b, Ns[-1]+1)   # Evenly spaced nodes on x-axis
6
7  X = np.linspace(a, b, 100)                    # points on x-axis for plots
8
9  plot_PolyAprox(f,  Ns, x_nodes, X, Pfit, Px, P_type=P_type+'Non_Poly')
```

$x \mapsto 0.6797623917902037\, P_0(x) - 3.0576890915194723\mathrm{e}{-15}\, P_1(x) + 0.9025518$
$456677498\, P_2(x) - 8.761237046307692\mathrm{e}{-16}\, P_3(x) - 0.5333448275283378\, P_4(x)$
$+ 5.510350741340122\mathrm{e}{-16}\, P_5(x) + 0.08770645807021914\, P_6(x) - 3.93030248$
$6420409\mathrm{e}{-17}\, P_7(x) - 0.003927370921187059\, P_8(x)$

It is found from Fig. 6.11 that for this non-polynomial function, the Legendre can only approximate it because it is not in the polynomial space. However, with the use of a higher degree of Legendre polynomials, the accuracy will improve. The convergence is very fast for this smooth function

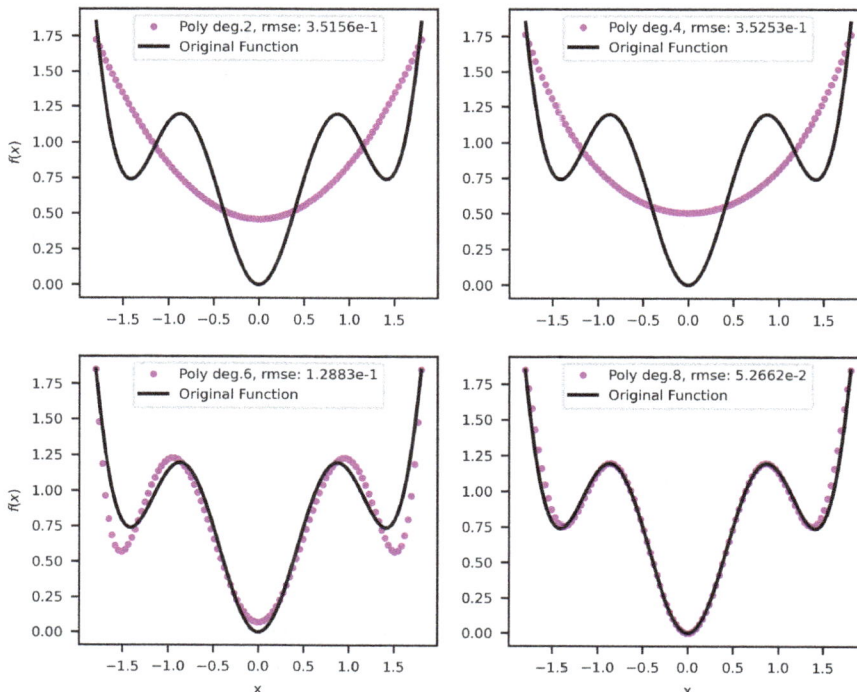

Figure 6.11. Approximated a non-polynomial function using the Legendre polynomials with different degrees.

(that is a composite of polynomial and sine function). This finding is also similar to what we found for the Chebyshev polynomials.

6.5 Use of Laguerre polynomials

6.5.1 *Formulation*

Laguerre polynomials also form the basis functions for approximating a function, as given in Chapter 3 and as rewritten here:

$$L_0(x) = 1 \quad \text{for } n = 0$$

$$L_1(x) = -x + 1 \quad \text{for } n = 1 \tag{6.13}$$

$$L_n(x) = \frac{1}{n}((2n - 1 - x)L_{n-1}(x) - (n - 1)L_{n-2}(x)) \quad \text{for } n > 1$$

The procedure to compute the fitting coefficients for Laguerre polynomials is the same as that for the Chebyshev polynomials. The difference is that the domain used for Laguerre polynomials is physical and is in $(0, \infty)$, and hence, no coordinate mapping is needed.

6.5.2 *Python example: Approximating a polynomial*

```
1  import numpy.polynomial as Poly
2  import numpy.polynomial.laguerre as Lag
3
4  f = lambda x: 2.1+5.7*x**2+1.5*x**3   # 3rd-deg. poly to be approximated
5  a, b = -1.8, 1.8                                          # Interval
6  Ns = [1, 2, 3, 4]                          # Highest degree of approximation
7
8  x_nodes = np.linspace(a, b, Ns[-1]+1)   # Evenly spaced nodes on x-axis
9                                                        # used for fitting
10 P_type = 'Laguerre'
11 Px = Poly.Laguerre
12 Pfit = Lag.lagfit                         # use help(Lag.lagfit) for details
13
14 X = np.linspace(a, b, 100)                   # points on x-axis for plots
15
16 plot_PolyAprox(f, Ns, x_nodes, X, Pfit, Px, P_type=P_type+'3rdP')
```

$x \mapsto 22.50000000000016\, L_0(x) - 49.80000000000057\, L_1(x) + 38.40000000000$
$094\, L_2(x) - 9.000000000000673\, L_3(x) + 1.709941810660917e\text{-}13\, L_4(x)$

As shown in Fig. 6.12, the findings are similar to what we found for the Chebyshev polynomials.

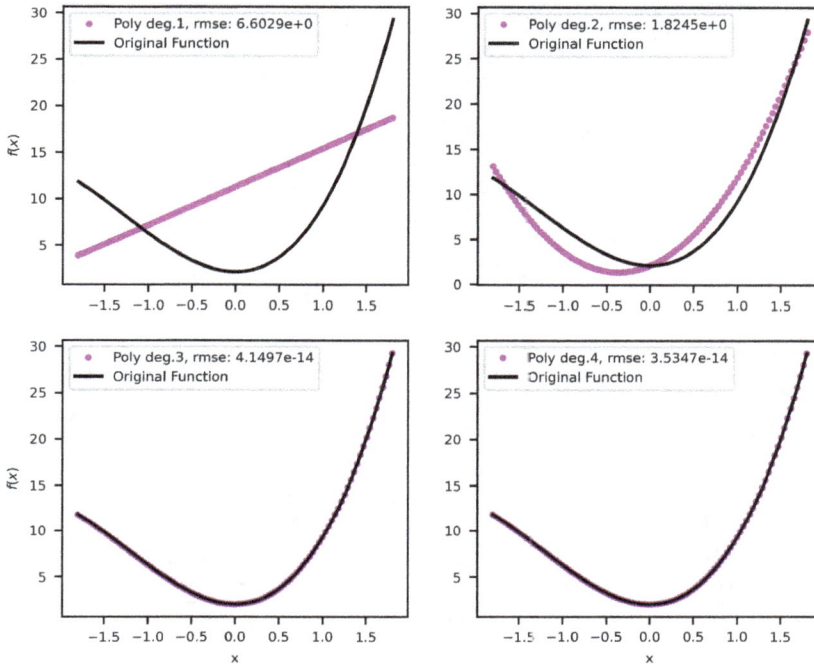

Figure 6.12. Approximated a third-order polynomial function using the Laguerre polynomials as basis functions with different degrees.

6.5.3 *Python example: Approximating an arbitrary function*

Let us approximate a general function that the Laguerre polynomial will not be able to reproduce:

```
1  f = lambda x: 0.2*x**3 + x*np.sin(3*x)    # Non-polynomial to approximate
2  a, b = -1.8, 1.8                                             # Interval
3  Ns = [2, 4, 6, 8]                          # Hignest degree of approximation
4
5  x_nodes = np.linspace(a, b, Ns[-1]+1)      # Evenly spaced nodes on x-axis
6
7  X = np.linspace(a, b, 100)                 # points on x-axis for plots
8
9  plot_PolyAprox(f, Ns, x_nodes, X, Pfit, Px, P_type=P_type+'Non_P')
```

$x \mapsto -6879.217167852167\, L_0(x) + 57019.31458255413\, L_1(x) - 205854.28079$
$387135\, L_2(x) + 422676.7815685646\, L_3(x) - 539708.0070733064\, L_4(x) +$
$438746.79675007804\, L_5(x) - 221727.54091261394\, L_6(x) + 63687.032053571$
$21\, L_7(x) - 7960.879007122785\, L_8(x)$

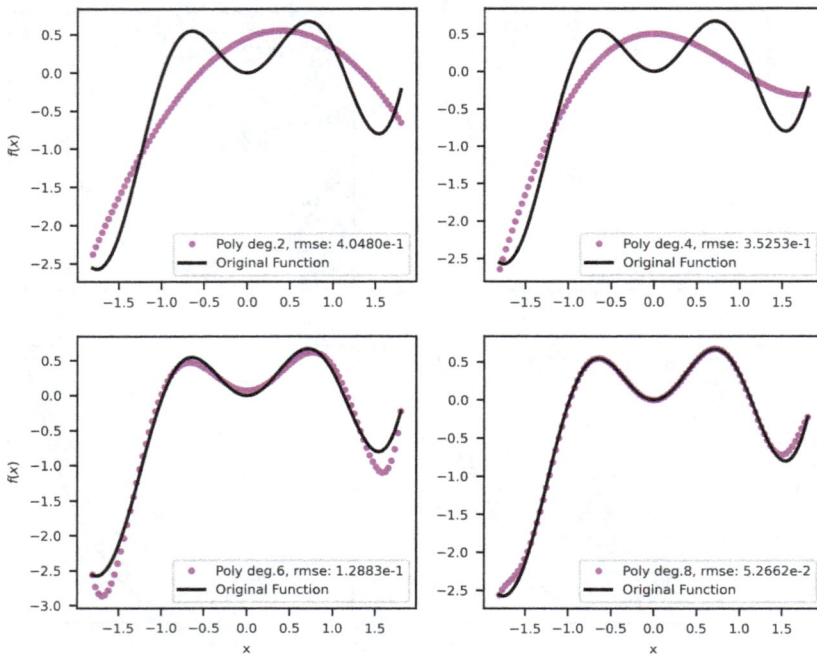

Figure 6.13. Approximated a non-polynomial function using the Laguerre polynomials as basis functions with different degrees.

The findings from Fig. 6.13 are also similar to what we found for the Chebyshev polynomials.

6.6 Use of Hermite polynomials

Fitting using the Hermite polynomial presented in Chapter 4 is done in the same way as the Laguerre fitting. The difference is that the domain used in the Hermite becomes $(-\infty, \infty)$, and hence, no coordinate mapping is needed. Here, we present a few examples, first, for approximating polynomial functions:

6.6.1 *Python example: Approximating a polynomial*

```
1  import numpy.polynomial as Poly
2  import numpy.polynomial.hermite as Her
3
4  f = lambda x: 2.1+5.7*x**2+1.5*x**3    # 3rd-deg. poly to be approximated
5  a, b = -1.8, 1.8                              # Interval
6  Ns = [1, 2, 3, 4]                      # Highest degree of approximation
```

```
 7
 8  x_nodes = np.linspace(a, b, Ns[-1]+1)      # Evenly spaced nodes on x-axis
 9
10  P_type = 'Hermite'
11  Px = Poly.Hermite
12  Pfit = Her.hermfit                          # use help(Her.hermfit) for details
13
14  X = np.linspace(a, b, 100)                  # points on x-axis for plots
15
16  plot_PolyAprox(f, Ns, x_nodes, X, Pfit, Px, P_type=P_type+'3rdP')
```

$x \mapsto 4.950000000000008\, H_0(x) + 1.1250000000000007\, H_1(x) + 1.4249999999$
$999978\, H_2(x) + 0.18750000000000022\, H_3(x) + 6.152578808260099\text{e-}16\, H_4(x)$

The findings from Fig. 6.14 are similar to what we found for the Laguerre and other polynomials studied earlier.

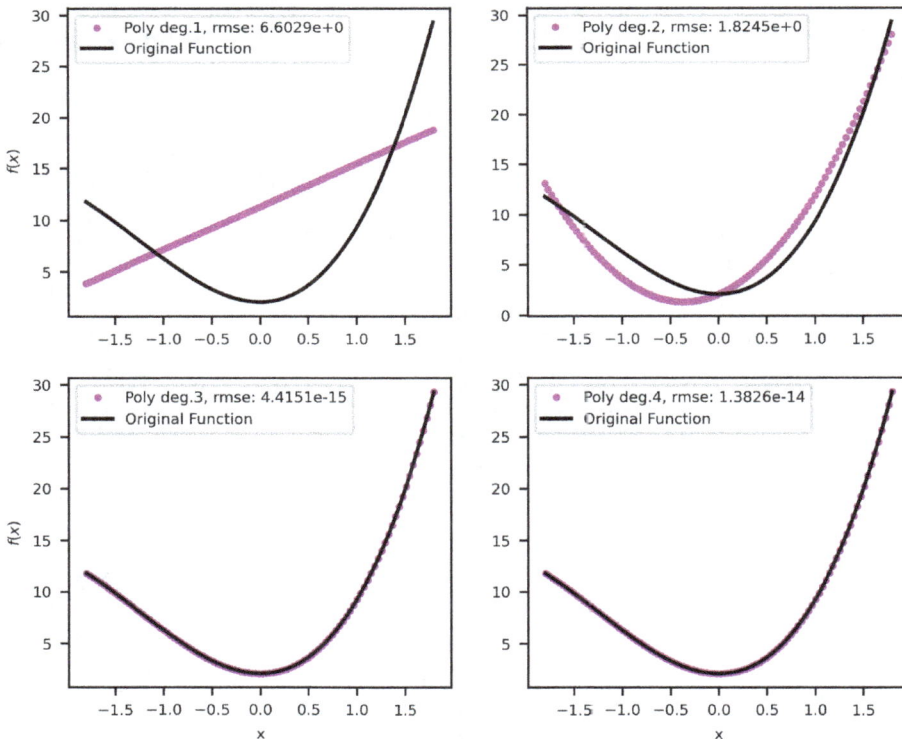

Figure 6.14. Approximated a third-order polynomial function using the Hermite polynomials as basis functions with different degrees.

6.6.2 *Python example: Approximating an arbitrary function*

```
1  f = lambda x: 0.2*x**3 + x*np.sin(3*x)   # Non-polynomial to approximate
2  a, b = -1.8, 1.8                                         # Interval
3  Ns = [2, 4, 6, 8]                             # Highest degree of approximation
4
5  x_nodes = np.linspace(a, b, Ns[-1]+1)     # Evenly spaced nodes on x-axis
6
7  X = np.linspace(a, b, 100)                    # points on x-axis for plots
8
9  plot_PolyAprox(f, Ns, x_nodes, X, Pfit, Px, P_type=P_type+'Non_Poly')
```

$x \mapsto 0.04907492398974173\, H_0(x) + 0.150000000000002\, H_1(x) - 0.456022327$
$8525764\, H_2(x) + 0.025000000000001653\, H_3(x) - 0.14858148275400598\, H_4(x)$
$+ 3.181347247984108e\text{-}16\, H_5(x) - 0.017646451693198173\, H_6(x) + 1.4733650$
$99677561e\text{-}17\, H_7(x) - 0.0007712595177563123\, H_8(x)$

The findings from Fig. 6.15 are similar to what we found for the Laguerre and other polynomials studied earlier. It is unable to reproduce the non-polynomial function.

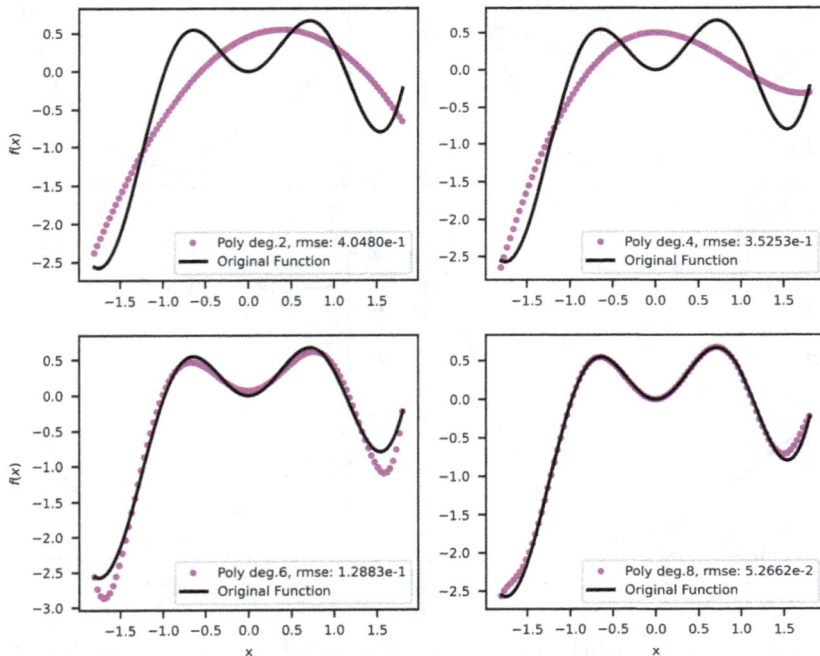

Figure 6.15. Approximated a non-polynomial function using the Hermite polynomials as basis functions with different degrees.

6.7 Use of shape functions

6.7.1 *Formulation and properties*

Using nodal shape functions created in Section 5.9.1, an arbitrary continuous function can be approximated as follows:

$$\langle f(x) \rangle = \sum_{i=1}^{N_n} f(x_i) N_i(x) \tag{6.14}$$

where $f(x_i)$ is the nodal value of the function at x_i, $\langle f(x) \rangle$ denotes the approximated function, and $N_i(x)$ is the nodal basis function for Node i. We shall note the following important features of this approximation:

- **Reproducibility at nodes:** Due to the Delta function property of $N_i(x)$, we found

$$\langle f(x_i) \rangle = \sum_{i=1}^{N_n} f(x_i) N_i(x_i) = f(x_i) \tag{6.15}$$

This means that the function values at the nodes will all be reproduced exactly.
- **Reproducibility of constants:** Assume that $f(x)$ is a constant: $f(x) = c$, where c is an arbitrary scalar constant independent of x. We now find

$$\langle f(x) \rangle = \sum_{i=1}^{N_n} f(x) N_i(x) = \sum_{i=1}^{N_n} c N_i(x) = c \underbrace{\sum_{i=1}^{N_n} N_i(x)}_{1} = c \tag{6.16}$$

This means that the constant function will be reproduced exactly. This is because of the partitions of unity property of $N_i(x)$ given in Eq. (5.26).
- **Reproducibility of linear functions:** Assume that $f(x)$ is a linear function: $f(x) = ax + b$, where a and b are arbitrary scalar constants independent of x. It will be exactly produced using Eq. (6.14). The proof is as follows. Using Eq. (6.14), we have

$$\langle f(x) \rangle = \sum_{i=1}^{N_n} f(x_i) N_i(x_i) = \sum_{i=1}^{N_n} (ax_i + b) N_i(x)$$

$$= a \underbrace{\sum_{i=1}^{N_n} x_i N_i(x)}_{x} + b \underbrace{\sum_{i=1}^{N_n} N_i(x)}_{1} = ax + b \tag{6.17}$$

This means that any linear function will be reproduced exactly. This is because of the partitions of unity property together with the linear reproducibility given in Section 5.9.1. It is like the Lagrangian interpolators: capturing the lower-order features for the function to be approximated.

The aforementioned proof is quite mathematical, but can be easily understood **intuitively**. The shape functions are linear in x, and they have the delta function property, implying that they pass through the two nodes (that form the element), between which x lies, therefore, any linear function there will be exactly produced, since two points determine uniquely a straight line. By this intuition, we can assert that if an element has three nodes, we can create three shape functions (using, for example, the Lagrange interpolators locally for the element) with the delta function property. Any quadratic function within the elements will be exactly produced. Further, this intuition can be extended to any order of approximations.

These properties of nodal shape functions as basis functions are similar to those for the Lagrange interpolators, but the domains of each basis function are different. Each of the Lagrange interpolator basis functions is over the entire $[a, b]$, as shown in first plot in Fig. 5.1. However, each nodal shape function basis is non-zero only in the vicinity of the node, as shown in Fig. 5.8.

6.7.2 *Python code for approximating a function*

We write a Python code to approximate a given function in a piecewise linear manner:

```python
def linear_approximate(x, x_nodes, f_vals):

    ''' Approximate a function  at x (scalar) piecewise linearly,
    using nodal base functions in natural coordinate.

    f_vals: given nodal function values at x_nodes'''

    # converting physical coordinate to natural coordinate
    ξ = (x-x_nodes[0])/(x_nodes[1]- x_nodes[0]  )         # for 1st node

    # function approximation using nodal shape function
    y = f_vals[0]*gr.node_shape_f(ξ)

    for i in range(1, len(x_nodes)-1):          # for nodes 2nd - Nn-1
        if x<x_nodes[i]:
            ξ = (x-x_nodes[i])/(x_nodes[i]  - x_nodes[i-1])
        else:
            ξ = (x-x_nodes[i])/(x_nodes[i+1]- x_nodes[i]  )
```

```
20          y += f_vals[i]*gr.node_shape_f(ξ)
21
22     y += f_vals[len(x_nodes)-1]*gr.node_shape_f(ξ-1)     # for last node
23     return y
```

The following code approximates a sine function using its values at a set of nodes:

```
 1  plt.rcParams.update({'font.size': 7})
 2  fig, ax = plt.subplots(1,1,figsize=(6,3))
 3
 4  # Define the x Locations, at which the function values are given.
 5  x_nodes = np.linspace(0, 2*np.pi, 11)
 6
 7  def f_true(x):                          # define a function
 8      return np.sin(x)+0.5
 9
10  f_vals = f_true(x_nodes)          # Set the corresponding function values
11
12  # Plot the original nodal fun. values and the approximated fun.
13  x = np.linspace(0, 2*np.pi, 101)
14  y = [linear_approximate(xi, x_nodes, f_vals) for xi in x]
15
16  ax.scatter(x_nodes, f_vals, c='r')      # original fun. values at nodes
17  ax.plot(x, y)                           # the approximated function
18
19  ax.grid(color='r', linestyle=':', linewidth=0.2)
20  ax.axvline(x=0, c="k", lw=0.6); ax.axhline(y=0, c="k", lw=0.4)
21  plt.savefig('images/linear4sine.png',dpi=500,bbox_inches='tight')
22  plt.show()
```

As seen from Fig. 6.16, the sine function is approximated in a piecewise linear manner using the linear nodal shape functions. The function values at

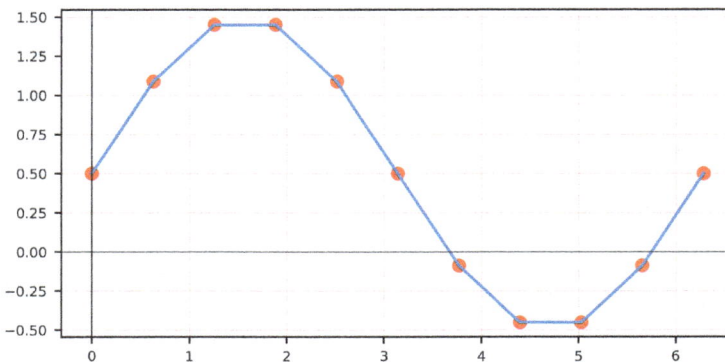

Figure 6.16. Approximated a sine function using linear nodal shape functions as basis functions.

these nodes are reproduced. One can use more nodes or higher-order nodal shape functions for better approximation between the nodes.

One may also simply use the built-in function interp1d() in Sympy to perform the interpolation, which also offers higher-order approximations. The following is an example:

```python
from scipy.interpolate import interp1d
fig, ax = plt.subplots(1,1,figsize=(6,3))

f_linear = interp1d(x_nodes, f_vals)                # linear interpolation

# higher-order interpolation:
f_quadratic = interp1d(x_nodes, f_vals, kind='quadratic')   # or 'cubic'

ax.scatter(x_nodes, f_vals,c='r')          # original fun. values at nodes
ax.plot(x, f_true(x), lw=3.5,    label='true function')
ax.plot(x, f_linear(x),          label='linear approximation')
ax.plot(x, f_quadratic(x), 'k:', label='quadratic approximation')

ax.grid(color='r', linestyle=':', linewidth=0.2)
ax.axvline(x=0, c="k", lw=0.6); ax.axhline(y=0, c="k", lw=0.4)
plt.savefig('images/linearQuad4sine.png',dpi=500,bbox_inches='tight')
plt.show()
```

As seen from Fig. 6.17, when quadratic approximation is used, the sine function is much better approximated.

Note that the accuracy and convergence of the approximation using nodal shape functions depend on the nodal spacing (hence the number of nodes used), which is known as h-convergence. This is different from that of using the Lagrange (and other polynomial basis functions), in which the accuracy and convergence depend on the order of the basis known as p-convergence.

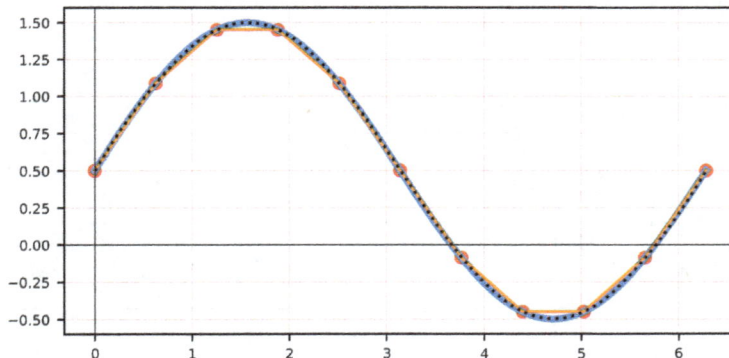

Figure 6.17. Approximated a sine function using linear and quadratic nodal shape functions as basis functions using sp.interp1d().

Readers may refine the aforementioned nodes and observe the accuracy and convergence of the approximation.

There are quite a number of ways to generate higher-order nodal shape functions for better approximations. The higher-order Lagrange interpolators are used in the finite element method [7]. A general point interpolation method is used in the smoothed finite element method [6] and the mesh-free methods [2]. In addition to polynomial basis, the so-called radial basis functions are used [2]. One can even create nodal shape functions using a series of neural processing units (NPUs), which is done in machine learning models [3].

Equation (6.14) is not only important for approximating a function with known values at nodes, as we did in the previous example. It is even more useful when the function itself is unknown and needs to be found out using other information on the function. When one tries to find a continuous function, it is often very difficult. With Eq. (6.14), the function is discretized and the unknowns become only its values at the nodes, which become finite. The variation of the function is taken care of by the nodal basis functions (that are known). This is essential in solving partial differential equations, as in the FEMs and meshfree methods.

6.8 Remarks

We finally mention some important remarks:

1. Basis functions can be used to form function spaces and then approximate a function using its values known at discrete points.
2. All the polynomial basis functions form a polynomial space. Any polynomial function with order lower than the order of the basis functions can be exactly reproduced. Functions that are not of polynomial type can be approximated.
3. We discussed the two approaches to perform the approximation: curve fitting and interpolation.
4. Interpolation can be done using the Lagrange interpolators in the whole domain. A more robust and scalable approach uses the nodal shape functions generated locally using the Lagrange interpolators. Linear independence is achieved by the distinct node locations of the nodal shape functions.

There are other classic techniques for approximation functions using the integral orthogonality of the polynomial basis, which is quite different from the curve fitting discussed in this chapter. For functions that are sinusoidal in nature, series expansion using trigonometric functions is useful and is known as the Fourier series expansion and Fourier transformation. All these techniques are better discussed after establishing the concepts of differentiation and integration of functions, which will be covered in a future volume of this book series.

There are quite a number of textbooks that use Fourier transformation for complicated problems. Interested readers may refer to Refs. [1, 8]. Reference [1] focuses more on classic theory and formulation, and Ref. [8] deals with advanced applications to elastic waves in composite laminated plates.

When we do not know the values of the function to be approximated, but know the differential equations and boundary conditions that control the function, we need a proper set of techniques to find the function in addition to the function approximation techniques discussed in this chapter. Interested readers may refer to Refs. [5–7].

References

[1] Liu GR, *Machine Learning with Python: Theory and Applications*, World Scientific, 2023.

[2] Liu GR, Dai KY, Han X, *et al.*, A mesh-free minimum length method for 2-D problems, *Computational Mechanics*, 38(6), 533–550, 2006.

[3] Liu GR, Quek SS, *The Finite Element Method: A Practical Course*, Butterworth-Heinemann, 2013.

[4] Liu GR, Nguyen TT, *Smoothed Finite Element Methods*, Taylor and Francis Group, New York, 2010.

[5] Liu GR, *Mesh Free Methods: Moving Beyond the Finite Element Method*, Taylor and Francis Group, New York, 2010.

[6] Liu GR, Liu MB, *Smoothed Particle Hydrodynamics: A Meshfree Particle Method*, World Scientific, 2003.

[7] James M, *An Introduction to Fourier Analysis and Generalised Functions*, 1959.

[8] Liu GR, Xi ZC, *Elastic Waves in Anisotropic Laminates*, 2001.

Index

www.ingramcontent.com/pod-product-compliance
Lightning Source LLC
Chambersburg PA
CBHW081515190326
41458CB00015B/5375